Funds, Flows and Time

Pere Mir-Artigues
Josep González-Calvet

Funds, Flows and Time

An Alternative Approach to the Microeconomic Analysis of Productive Activities

With 73 Figures and 8 Tables

Springer

Dr. Pere Mir-Artigues
Department of Applied Economics
University of Lleida
Jaume II, 73
25001 Lleida
Spain
peremir@econap.udl.cat

Dr. Josep González-Calvet
Department of Economic Theory
University of Barcelona
Av. Diagonal, 690
08028 Barcelona
Spain
jgonzalezcal@ub.edu

ISBN 978-3-642-09037-0 e-ISBN 978-3-540-71291-6

Springer is a part of Springer Science+Business Media

springer.com

Cover-design: WMX Design GmbH, Heidelberg

Foreword

The subject of this book is *production*, which is an important and extensive field in economic science. In fact, *production*, *distribution* and *consumption* were long considered the three federated kingdoms which together formed the great empire of the economy. According to other slightly different traditions, production also held pride of place, specifically as a basic link in the long chain of social reproduction. Today, whatever the theoretical approach, production is a fundamental requirement for human survival. This was not, however, always the case. For much of the history of mankind hominids were hunter, scavenger and gatherers, with very little control over their environment, and extremely little in the way of artefacts with which to work. However, since the Neolithic revolution, productive processes have constituted an essential mechanism, providing human society with goods and services to satisfy its needs and cravings.

A simple, yet pertinent, characterisation of the *production process* conceives it as the transformation of a conglomerate of factors into a given number of products within a specific period of time. Refining this definition a little further, the said factors may be broken down into different categories: natural resources, means of production (covering two species: working capital and fixed assets) and the different forms of specific work. To the above, we must then add a minimal economic environment consisting of suppliers, clients and a legal and social framework. All of these elements act upon each other, are directed by intentional or routine behaviour, and are sustained by organisational structures which, to a greater or lesser extent, combine principles of co-operation and hierarchy.

The paradigmatic notion with which this metamorphosis is generally described is what is commonly known as the *production function*. This term may be used to refer to two objects of very different dimensions and clearly distinct transforming connections. On the one hand, we have microeconomic production functions with, at least in principle, perfectly well identified points of reference and clearly defined relationships. On the other, we have aggregate (or macroeconomic) production functions which link virtual amounts of capital and work to a specific amount of aggregate production by way of an ethereal and purely formal connection. In this respect it should be emphasised that microeconomic production functions are irreproachable theoretical constructs with a clearly defined correlation, while the presumed functions of macroeconomic production are a very different matter altogether. The latter lack any real counterpart and thus appear to be eternally condemned to explicatory and analytical opacity. It is not feasible to quantify their value independently, nor do they appear to contain any underlying aspects or mechanisms that are waiting to be revealed by theoretical or practical advances in scientific research (as occurred with genes, atoms or the subconscious mind). In short, as they lack explicatory force they may only give rise to spurious correlations, with a predictive capacity which goes no further than any astutely constructed econometric extrapolation.

This book by Professor Pere Mir-Artigues and Professor Josep González-Calvet exclusively addresses this hard core of microeconomic production functions. In my opinion, the main merit of the book lies in its painstaking theoretical work, which has been built to embed, within this conceptual artefact, certain basic aspects related to time, a dimension that it is not at all easy to represented. For example, in few fields relating to the economy is time so vital as in the area of production. It is therefore not unusual for production functions to be presented as analytical schemes bereft of this very crucial dimension. Another unforgivable way of overcoming such obstacles is to postulate *instant* production functions, thus directly opposing the basic laws of physics. I maintain that one must distinguish between simplification and idealisation on the one hand and absurdities on the other. The boundary between the two is somewhat difficult to establish but, in general, the specific cases dealt with by economists clearly pertain to one or the other of the above categories. Let it therefore be accepted that, while it may seem perfectly licit to assume that two subjects have the same tastes, or that two production processes take the same time, it does not seem acceptable to postulate that people are immortal, that agents are omniscient, or that production processes are instantaneous. In short, such outlandish postulation is not acceptable, unless one wishes to entertain oneself with academic games or formal virtuosity, rather than face the reality of the world.

Notwithstanding, there are forms of idealisation, or borderline cases, which may prove illustrative. An example of these is the *stationary state* device, which permits the presence of change and time, but minimises the farrago of complications that surround such aspects. Apropos of which, I wish to state that it is a (frequently made) mistake to consider the terms static and stationary state to be synonyms. To set the record straight, it is sufficient to realise that with the term static, time does not exist (if anything, there is just calendar time, but there is never any idea of duration) while, on the other hand, time is present in the ideal trajectory of a stationary state. But then time flows without modifying structure, so the core variables have constant values given that any potential changes have been sterilised and reduced to a minimum. Let me clarify this with a simple illustration: a snapshot (a typically static form of representation) of a person does not, in general, provide information on what the individual had for lunch that day. However, if we were to imagine that his/her weight has stayed the same over the last month, then it is logical to infer that from day to day, in terms of calorie intake, his/her diet could be defined with precision.

Apart from the aspect of time, it should be stressed that relationships between the factors and product, and between the factors themselves, are more varied than the format customarily applied to the production function would tend to suggest. To be more specific, standard economic theory emphasises the interchangeability of factors and the impact that marginal variations in each factor may have on total product volume. Both attributes are often baseless assumptions. In fact, there is little room for manoeuvre once the production plan has been designed and the pertinent machinery installed. Moreover, there is no lack of discontinuity or indivisibility, and very often complementarity prevails over interchangeability. In pedagogical terms, greater emphasis should be placed firstly on treating the factors of production as discrete variables, secondly on lineal programming as a very valuable technological tool then, finally, on the analytical treatment of the different temporariness involved in production processes. Of course we must rejoice in the tremendous advances made over recent years in terms of transaction costs and information, but nonetheless a systematic general effort should be made to reveal the basic mechanisms at work, bearing in mind the real physical, chemical, biological, engineering and organisational characteristics underlying the different production processes.

It is not the aim of the authors to cover such broad terrain. Their goal is, however, to provide precise, accurate results with respect to the treatment of time, funds and flows. The book is complex and yet simple. It makes an outstanding contribution to the analysis of production processes from an original and potent perspective. The authors have exhaustively combined two highly valuable aspects, neither of which would normally be expected in a study of economic theory: representational realism and practical applicability. Their specific object is to elucidate how, in terms of time, the different factors of production are articulated. In other words, it examines the *services of human labour* and the two main groups into which the means of production are divided: the means of production that operate as working capital (*flows*) and those that operate as fixed assets (*funds*). The book examines this without any collateral inquiry. It might well seem that economic processes are merely conditioned by inherited technology, while the social structure, legal and political framework, financial institutions, habitat, climate, gender relations and type of market in which suppliers and clients negotiate are unimportant. There is no doubt that the authors are well aware of the fact that no social relationship is completely alien to the context that surrounds and shapes it, but it is always recommendable to make reference to this.

The authors approach their subject matter following the directives proposed by Nicholas Georgescu-Roegen (1906-1994), a heterodox guiding light of 20th century economic thought. He stands out for his ecological sensibility, his criticism of the utility function and production function, his ceaseless work on analysing formal fundamentals and tools, and his constant exactingness and realism. The approach adopted by Mir-Artigues and González-Calvet is built around two pivotal points: on the one hand, the explicit, deliberate consideration of the dimension of time, which is an essential, unavoidable, feature of all and any production process and, on the other, their aspiration to reconcile representational realism and practical applicability. Nonetheless, it is not always easy to surmount such obstacles while also avoiding invalid parallelisms. For example: if a sow has 10 piglets every 10 months it is not the same as if she had one piglet every month for 10 months. It is not the same in real or economic terms, and, if it were said to be the same, solid proof of this would need to be presented. There should be no need for such caution but in reality, it is better to be safe than sorry, so this is not a wasted exercise.

In a different light, it should be noted that the subject of production is, in general, presented with an ambiguous epistemological statute and straddles the disciplines of science and technology. Of course there are, have been, and always will be great fertile bonds between the corresponding branches of pure and applied science in any discipline. But it is not a good idea to blend or blur the boundaries between the two (or more) planes. Such ambivalence occurs in the area of production given that economic theory most particularly concerns itself with suggesting and illuminating forms of action for the economic agents and with far less interest in how the world really works. Such ambivalence is also present in this book given that the schemes proposed simultaneously attempt to satisfy both a representational and a pragmatic aim: they are both legitimate goals, but objectives requiring clear delimitation

Alfons Barceló
University of Barcelona
February 2007

Preface and acknowledgements

This book deals with many questions concerning production economics using the fund-flow model of Nicholas Georgescu-Roegen as a basic reference. Despite the fact that this proposal is currently considered rather heterodox, there is no difficulty in structuring the text along the lines of most other manuals of microeconomic production theory. However, there are some important differences to be taken into consideration.

We start by outlining the basic concepts which the model is built around, i.e., the notion of funds and flows and, in particular, that of the elementary process. These tools enable a highly detailed representation of productive operations to be developed that is open to partial modification when applied. Furthermore, the seminal concepts proposed by Georgescu-Roegen make it possible to shed new light on certain long-established concepts of microeconomic production theory as well as to open the way towards an original analysis of the organization of productive processes.

The first chapter also deals with certain assumptions that limit the analytical scope of the model presented. These are problems that tend to be shared with all other forms of partial interdependence or equilibrium approaches. These limits do not suppose a complete loss of relevance by the model, but simply call for due care with respect to its analytical scope.

Three templates for the organisation of production are explained. These represent the generic ways in which it is possible to manufacture one unit of output after another for an unspecified period of time. At this point, the book follows the triple classification proposed by Georgescu-Roegen, adding only a few minor refinements which stem from the work of other authors interested in his analytical-descriptive approach, along with some of the results obtained in the process of using the strategic manufacturing theory. In doing so, Georgescu-Roegen's proposals could be linked to some of the fundamental results of management science.

An extensive chapter analyses the peculiarities of the production in line. This central form of productive organization is dealt with by using the model in a very detailed manner, which -in turn- leads on to an exhaustive discussion of process timing optimisation. As we know, operational research has developed many practical tools in its pursuit of the elusive goal of timing optimisation. A link could therefore be established between the fund-flow model and research into optimisation. This is perhaps the main difference between this book and most other texts on Microeconomics.

Analysing all the aspects of the process in line also implies a reinterpretation of the well-known concepts of scale and indivisibility. The funds and flows model takes a new look at these concepts and seeks to redefine and highlight new aspects of them.

Although the fund-flow model was developed to represent production processes with tangible outputs, there is nothing to prevent an attempt to extend its application to other forms of activity. An entire chapter has therefore been dedicated to a preliminary investigation of the application of Georgescu-Roegen's ideas to what are known as service activities, making special reference to transport operations and the development of information assets. However, since tertiary activities constitute a highly diverse and ever-changing phenomenon, attempting to cover all the idiosyncrasies therein in any depth would require a study that goes far beyond the scope of this book.

As the reader might expect, the book finishes with the topic of costs. Economists are currently concerned with the way in which costs react to changes in the level of production. Here, however, the analysis is somewhat different: it is a question of ascertaining the behaviour of costs over time, and particularly of analysing their behaviour in the light of the deployment of productive activity. This inclusion of the time factor in production cost analysis sheds new light on issues such as the impact on costs of changing the length of the working day, whether by means of extra hours or extra shifts. Of course, it is also only natural to take any associated costs relating to transport services and software products into consideration.

Although this study might be used as a textbook, particularly for advanced courses, it must be acknowledged that this book is no more than an introduction to the subject dealt with. Indeed, the aim of the following pages is to establish certain fundamental concepts of microeconomic production theory (such as scale, indivisibility, or flexibility), working with the analytical fecundity of the ideas proposed by Georgescu-Roegen. It includes concepts that, despite having mainly been presented for the first time in the 1960s and 1970s, have yet to receive all the attention that they deserve. Since the following pages only offer a sample of the different analytical developments that may arise from the application of a funds and flows approach, readers are invited to further develop the model in question themselves, albeit empirically, or just for the pure academic delight of doing so. It must be said that with respect to the microeconomic analysis of production, the richest fruit from the seeds first planted by Georgescu-Roegen have yet to be harvested.

In the chapter of acknowledgements, we would like to thank Ch. Boswell for text revision. We would also like to thank the following for their valuable comments and observations, some of which were made during the writing of the book while others were reactions to shorter preliminary studies: A. Barceló, G. Cortés, M. A. Gil, M. Morroni, E. Oroval, A. Petitbò, P. Piacentini, A. Recio and J. Roca. Two anonymous referees also made fitting and valuable suggestions, most of which have been incorporated into this final version. It goes without saying that the authors alone are responsible for any errors or omissions the text may still contain.

Pere Mir-Artigues and Josep González-Calvet
Lleida
February 2007

Contents

List of most important symbols

$[0, T_p]$	*Duration o the productive process*
J	*Duration of the working day*
$\boldsymbol{N}_{j,s}$	*Sub-table of time service of fund j in phase s*
$\boldsymbol{A}_{k,s}$	*Sub-table of inflow/outflow k in phase s*
$\boldsymbol{P}_{j+k,s}$	*Table of the elementary process*
$I_k(t)$	*Function of the amount of k element entering the process*
$O_k(t)$	*Function of the amount of k element exiting the process*
$F(t)$	*Accumulative functions measuring flows*
$U_j(t)$	*Functions measuring j fund activity*
$S_j(t)$	*Total accumulated services rendered by the j fund*
$[0, T]=T$	*Duration of the elementary process*
δ	*Maximum common divisor of the duration of the stages of an elementary process*
δ^*	*Maximum time lag for the optimum functioning of a given line process*
S_p	*Size of the process in line*
c^*	*Common cycle time of a balanced in line process*
p_i	*Transformation work time*
π_i	*Non-transformation work time*
e_i	*Idle time*
κ^*_j	*Maximum number of elementary processes that j-type fund may simultaneously process*
γ_j	*Index of used capacity of j-type multiple capacity fund*
τ_j	*Minimum common multiple of a coordinated multiple capacity funds*
H	*Total working time per annum*
W	*Annual payments for the services of human work*
σ_j	*Annual cost allocation for the use of j funds*
f_k	*Technical coefficients of the flows entering the process*
p_k	*Exogenous prices for flows*
D	*Number of work shifts in a calendar year*
Γ	*Cost of the first copy of an information product*

For Imma and Remei

1 Anatomy of the production process

1.1 Economic models of productive activities

From the very outset Political Economy has always had a particular interest in production. Nor indeed could it have been any different as humankind has always been eager to ensure its material needs are covered and guarantee its physical survival, reproduction and the fulfilling of the different needs it has developed over the course of history. This has resulted in an infinite variety of actions aimed at gathering, storing and processing all sorts of tangible goods, as well as social or personal services. This extensive set of intentional processes is generally known as *production*. It is a fundamental economic activity, the execution of which has always required different amounts of resources and time and over the course of history, this has involved the development of highly diverse social relations and technological skills.

The term production covers an extensive semantic field. In this book it is understood to mean the process of creating value by way of the coordinated execution of a given number of operations over a specific interval of time. Therefore, the purpose of production is:

1. To change the physical and/or chemical characteristics of different types of material or objects.
2. To extract natural resources for consumption or processing, or to harness certain natural resources capable of generating energy, by such means as water mills or hydroelectric power stations.
3. To systematically exploit the reproductive potential of plants or animals to obtain a highly diverse range of goods.
4. To provide different kind of services. As is known, transport operations, commercial distribution or telecommunications services are added value activities because a quantity of resources are needed to provide them. As is also known, pure services do not exist.[1]

---

[1] It should be pointed out that services have been traditionally excluded from production models, i.e., output was implicitly understood as a *tangible* good. However, in this work, the main attribute of production activities is considered

The fact that Political Economy has taken productive activity as one of its main fields of study accounts for the proliferation of models developed with respect to it. Knowing that a model is a rational reconstruction of those features considered essential parts of a given phenomenon, the economic analysis of productive processes has developed a series of very different theoretical approaches. Depending on the case in question, proposals have emphasised certain ingredients (such as materials, skills or functions) and some aspects to the detriment of others. The diversity of methodological and theoretical approaches therefore makes it difficult to achieve any kind of true systemisation. However, a degree of order must be established for any progress to be made.

Firstly, it is essential to establish the main focuses under which the phenomenon of production has been examined up to now. To date, the subject of production has largely been approached by adopting the following three points of view:

1. The transformation of given inputs (or elements entering the production process) into tangible outputs (the results of the process of transformation). This approach focuses on the technical features of the process and on the different elements involved in it. The description of productive operations, be they mechanical, chemical, or whatever, is the main object of analysis. Thus the ultimate object of any such investigation is to perform a detail dissection of the production process.
2. The creation of value. There is no doubt that this is the most genuine approach in terms of economic analysis. Initially such models are built around assumptions that, to a greater or lesser extent, reduce the organisational complexity of the production process. By incorporating prices and certain distributive variables, it is possible to calculate how the net value created is distributed between the main players involved in the process. In doing so, it is possible to explain the mechanism by which the product is appropriated.
3. This process is understood as an articulated set of decisions, running from the design stage to the control of manufacturing and including, of course, the detailed planning of productive operations. Here the emphasis is placed on the methods employed to achieve certain technical and/or financial goals considered desirable by management. To such ends, the result of the production process is measured by way of generally accepted indicators.

to be their added value. Therefore services are also analysed through the fund-flow model. Obviously, when dealing with other models of production processes the assumption that output is a tangible good is maintained.

With the exception of the third of the above approaches, which has a strictly pragmatic nature, analytical models have always combined aspects of material transformation with the creation of value. Given the enormous variety of such models, a possible criterion for classification could be the plane of analysis within which they are defined; in other words, models of general or partial interdependence.

The first type of model comprises circular aggregate schemes, the earliest versions of which were formulated by the physiocratic authors, along with the authors of the Classical Political Economy. They build upon the sectoral interdependence that exists in all economic systems and go on to analyse the conditions for reproduction, and for the distribution of the surplus generated. Such models adopt a long-term outlook.[2]

On the other hand, other models opt for the partial equilibrium approach. In this case, most of the reciprocal influences that exist between the different sectors of production are left aside and an articulated set of hypothesis relating to the *internal* configuration of production processes is developed (Tani 1989, 1993). Even though a degree of influence is afforded to the most immediate environment (the input and output markets) its role is reduced to a simple data. Or, put in other words, factors behind the decisions made by the enterprises involved are not taken into account. Moreover, in line with the great constraints within which the process of production is analysed, a one-way view of the latter is preferred: only the *direct* relationship, from raw material to goods produced, is considered. Despite the undeniable advantages such a restricted outlook offers when tackling, in great detail if desired, questions relating to the organisation of the process of production, given their contextual shortcomings, the use of such models of partial equilibrium is risky.

Bearing in mind the cost entailed in any form of classification, models of partial interdependence or equilibrium may be divided into two main types, the most widespread of which has a clearly normative goal: the pursuit and refinement of the principles of optimisation. This group comprises the numerous formal variants of the so-called neo-classical function of production (Shephard 1970; Frisch 1965; Ferguson 1969; Johansen 1972; Fuss and McFadden 1978) as well as some theoretical proposals, which are disappointing for certain of their assumptions. Outstanding amongst the latter are the engineering production functions, i.e., models with mathematical framework and data supplied by technical projects (Chenery 1949;

---

2 There is no doubt that the Sraffian proposal (Sraffa 1960; Kurz and Salvadori 1995) is the best developed of the general interdependence models. The works by Quesnay and Marx also belong to the same group.

Tinbergen 1985; Whitehead 1990)[3] and the lineal activity analysis (Koopmans 1951).

The second type of model includes those that seek to offer a detailed description of the production process. On the one hand, the nature of the materials used, the type of operation performed and the skills required from those performing them are given preferential treatment. In doing so, a substantial part of the classical discourse on the effects of the technical division of labour and the mechanisation of the production process is recuperated. On the other, in addition to classifying the elements and functions of the production process, special relevance is attached to the time dimension. Since there are some differences, two models could be considered in this approach: the fund-flow model, called the *analytical-descriptive* method by its inventor Georgescu-Roegen[4], and the model that considers the production process as a network of tasks (Scazzieri 1993; Landesmann and Scazzieri 1996a, 1996b). Nonetheless, a convergence between the two approaches is acknowledged by the authors of the latter, who confess their intellectual debt to the former, initially proposed almost three decades earlier.[5]

Before ending, it must be pointed out that our brief classification has not included the (neo)Austrian approach (Hicks 1973; Amendola and Gaffard 1988). This approach considers the production process as a compact, continuous, one-way current of materials that becomes, with the passage of time, a flow of products finally to be consumed. Two flows that vary over time are measured in terms of money. According to this model, fixed as-

---

[3] Engineering models of the production function express technical relationships using one or more explicit equations. This type of production functions are built when the process of transformation is reasonably homogenous and continuous. For instance, river cargo transport (Ton-Km/h) as a function of engine horsepower and the size of the boat, or steam generation of electricity (Kwh) expressed in terms of boiler temperature and pressure (De Neufville, 1990). Engineering production functions entail an important effort of data collection from experimental results. The identification of the production frontier is achieved by estimating several production functions.

[4] Unfortunately, nowadays the life and work of Nicolae Georgescu-Roegen (1906-199 4) is very little known amongst economists. References to him are to be found in Zamagni (1982), Georgescu-Roegen (1989, 1994), Maneschi and Zamagni (1997) and Bonaiuti (2001).

[5] It was at the *Conference of the International Economic Association*, held in Rome in 1965, that Georgescu-Roegen first presented the complete model of funds and flows (Georgescu-Roegen 1969). Since then the model has appeared, with no appreciable changes, time after time in his work. Notwithstanding, the last study wholly dedicated to this model was published in the mid-seventies.

sets are considered mere interim products. Despite the importance attached to time, this approach completely ignores the internal structure of the production process.

A general chart of the different economic outlooks on production is shown in figure 1, with special emphasis placed on models of partial equilibrium.

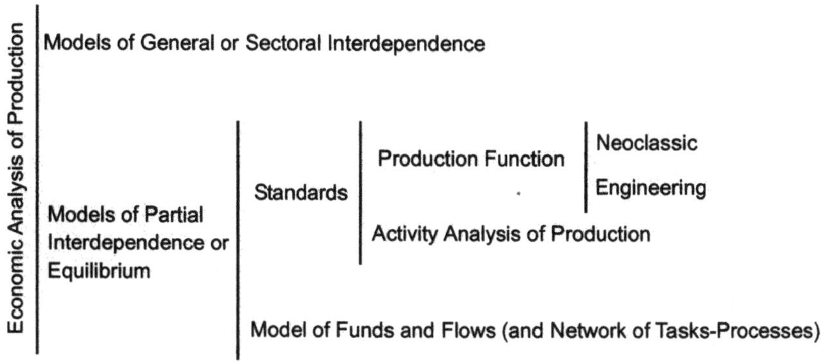

Fig. 1. Economic models on production

The theoretical outlooks have been deployed in innumerable models, the majority of which are eminently mathematical. This accounts for the quantitative nature of the information processed which, it must be said, has never been difficult to obtain. Therefore, there is no surprise about the proliferation of this kind of publication, especially those relating to the postulates of the most common production function model. So widespread in fact that it is considered by many to be the conventional model.

This book is based on the model initially formulated by Georgescu-Roegen, and the further development suggested by some of those that followed him. These authors are fully convinced that Georgescu-Rogen's proposal is the most suitable line of research for tackling the phenomenon of the internal organisation of the production process and to represent, with a satisfactory degree of detail, the key aspects of the process technical changes. Thus the analysis of the production, unfortunately hidden in the *black box* normally used, is decisively improved. Nevertheless, certain authors scorn this approach as they consider such detailed models to pertain more to the domain of engineering (agricultural or industrial) than economics. This criticism, as the reader will observe, is not acceptable. Indeed, once the black box has been opened, economic science may still offer great contributions to the understanding of the production processes. Therefore, it may be suspected that that argument is, in truth, a pretext to avoid undertaking a general revision of the theoretical guidelines received.

Instead of refining and replacing the current conceptual tools, it is easier to declare that certain issues are out of the sphere of the economic analysis. Despite this, the existence of inevitably fuzzy boundaries and the need for interdisciplinary co-operation do not justify completely washing one's hands of a promising field of research.

Production management researchers have always given maximum priority to subjects such as task organisation or the time balancing of the manufacturing processes and the repercussions on costs. To these ends, they have developed a broad array of models, some of which are highly sophisticated, without practically considering the postulates of the microeconomic theory of production. Indeed, most of the conceptual devices do not serve for their purposes. Thus, the present book is an attempt to close the unjustifiable breach that exists between economic theory and the more pragmatic approaches to the production process. Although the territory of each discipline should be respected, it is unforgivable not to establish bridges between them. Moreover it is unacceptable for conventional theory not to design models in which time, a key dimension in any production process, does not take pride of place.

1.2 Basics of the funds and flows model

First of all it is essential to establish a few of the general properties of productive activity which, along with other more specific assumptions, dealt with later on, form the core of the model of funds and flows. This set of propositions is shared with many, although not all, of the other approaches to production (adapted from Lager 2000: 233):

1. Production requires time. There is no such thing as instantaneous production.
2. Prior to starting a production process inputs are needed.
3. Any transformation or processing of inputs generates one or more outputs. The physical nature and economic value of these outputs may differ greatly but, nonetheless, will always exist. Indeed, the end result of a production process may never be nothing at all.
4. At any given moment in time there will always be a finite number of methods of production.
5. There is a finite number of inputs and outputs. Despite their material differences, position in space and availability over time, they shared some characteristics that contribute to their classification.

Having established the above principles, from the methodological point of view a detailed description of productive activities presents two major aims:

1. To establish the boundary of the process. This separates the process from anything else. According to Georgescu-Roegen (1971: 213, underlined in the original),

   No analytical boundary, no analytical process. (...) where to draw the *analytical* boundary of a partial process –briefly, of a process- is not a simple problem. (...) a relevant analytical process cannot be divorced from purpose.

   There are two types of boundaries: the frontier which separates, in analytical terms, the process considered from its environment at any point of time, and the duration of the process. It must be said that it is always possible to fix more boundaries in order to divide what was a single process into two o more processes.
2. To draw up an exhaustive list of the co-ordinated elements participate therein.

Time limits permit, on the one hand, the process itself to be recognised and, on the other, to order and date its different constitutive stages. With respect to cataloguing the elements of the process, what is directly apparent is that the process has inputs and outputs. Thus at instant t of the process, $t \in [0, T_p]$, one or more of the elements will enter or leave the process, as shown in figure 2.

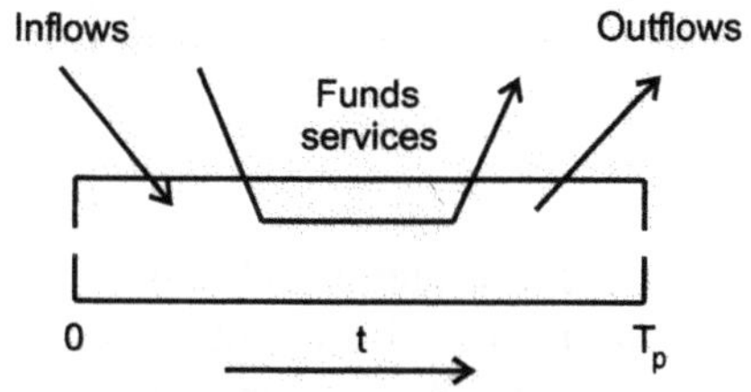

Fig. 2. The production process

The figure shows a scheme of the confines of a production process which begins at 0 and ends at T_p. It also shows the one-way nature of the time dimension.[6]

[6] Time is herein understood in the Newtonian sense. This meaning is enough for our purpose: time is the continuous, homogeneous, and one-way displacement from an irrevocable past to an uncertain future.

1.2.1 Funds and flows

Many different elements are involved in the production process. One possible classification of these would be based on their behaviour with respect to the boundaries of the process of transformation. Thus, as an initial criterion, dividing the elements into funds and flows is proposed:

1. The funds furnish certain services over a given period of time. Thus they enter *and* leave the process.
2. The flows enter (inflows) *or* emerge from (outflows) the process.

Both funds and flows are cardinally measured.[7] Four different types of funds may be distinguished:

1. The different kinds of workers whose services (also known as *human work*) form an essential part of all necessary, intentional transformation activity.
2. The land, taken in the sense established by Ricardo, that is, as the area on which productive operations take place and also the surface which sunlight strikes and is captured by.
3. The means of production: either previously manufactured equipment (such as tools or machinery) or constructions (such as premises or goods warehouses), living beings that participate actively in the process (such as fruit trees or draught animals) or are structurally required as facilities (such as trees acting as windbreaks).
4. The populations of natural organisms integrated into the ecosystem that, with their services, co-operate in the productive activity. Such, for example, is the role of bees in the pollination of crops.
5. The *process-fund*. This is comprised of the overall units of output *still* in process.[8]

Funds are not *physically* incorporated into the product. Moreover, with the exception of the *Ricardian* land, the operative capacities of the human work fund and the means of production are reduced after the process. At the end of the day, the workers are tired and with use, the fixed capital as-

7 The classification of funds and flows established by Georgescu-Roegen (1969, 1971, 1990: 207ff.), is repeated by authors such as Ziliotti (1979: 628-630), Tani (1986: 200, 1993) and Scazzieri (1993: 110-112). In the text it has been added the funds encompassed by the term *natural capital*.

8 It must be said that the inclusion of this fund has been criticised as its role is strictly passive (Landesmann and Scazzieri 1996b: 222). Notwithstanding, in these pages to avoid a problem of completeness and consistency in the analytical representation of the production in line this fund will be considered. See chapter 4.2 below.

sets suffer wear and tear. In general, the labour and machinery funds require an appropriate flow of energy and material to preserve their productive capacities at an acceptable level although, over the passage of time, their capacity is progressively and inevitably reduced. Over the years such funds age. For the sake of simplicity the operations aimed at restoring, maintaining and repairing them could be considered as processes separate to the main process under examination (Georgescu-Roegen 1969, 1976, 1990). At this point, the funds of workers and capital equipment will be distinguished (Georgescu-Roegen 1971: 229ff.).

Flows may be divided into five categories:

1. Goods produced by previous processes, that is, raw materials, unfinished products and components, seeds and energy.
2. Natural resources at a positive or zero prices: sunlight, air, water, minerals, and so on.
3. The main output (one or more products) obtained at the end, or at an intermediate phase of the production process.
4. Output produced below the quality standards established at the time by the company.
5. Waste and other residual products and emissions (gases, particles, radiation, etc.).

The first two types of flow enter the process and are incorporated into the different products that result from it. On the contrary, the three types of output leave the transformation process without ever having entered it.

Although we have considered two main types of factors of production, and given just a few examples of these, it does not mean that they could not form unlimited different specific configurations, or be involved in countless technical combinations of greater or lesser complexity.

Having defined and classified the elements of production, our analysis must now turn to the time dimension of the production process. So, as has already been indicated, it is essential to precisely establish the instant of starting (0) and finishing (T_p). Thus, all processes are of a specific duration $[0, T_p]$. Within the aforementioned interval, as one would expect, the activity of each of the different co-ordinated elements has its own particular trajectory through time.

As has been suggested, how the boundaries are precisely set is a decision that will depend on the objectives of the research. In general, activities performed to sustain the productive capacity of the funds involved in the process, and those related to the securing of materials and the distribution of products, tend to be considered as separate processes. This means a loss of generality, but simplifies the analysis. Nonetheless, this assumption may be forsaken without affecting the consistency of the model in the slightest.

1.2.2 Self-reproductive goods

Discriminating the elements of production in funds and flows is a purely functional exercise, i.e., the same asset may be classified differently according to the role it plays in the production process: a new tractor fresh off the production line would be considered a flow, while once working on a farm it should be understood as a fund. There are innumerable examples of this type. Notwithstanding, Georgescu-Roegen states that:

the clover seed, in a process the *purpose* of which is to produce clover seed, is a fund, but in a process aimed at producing clover fodder, it is a flow (Georgescu-Roegen 1969: 511, cursive in the original).

On careful examination, this statement reveals a problem with respect to the classification of self-reproductive goods.

Animals and plants represent the bulk of self-reproductive goods. This is a productive area in which the main technical and economic interest lies in the different rates of reproduction which may be calculated by properly comparing output and input volumes (Barceló and Sánchez 1988: 14-15). There is no question that the management of the reproduction and growth cycles of such goods has constituted a fundamental economic activity of all human society since time immemorial.

Could it not be considered that the fund-flow classification would not hold valid for such self-reproductive goods given that the main output and input are often the same. In effect, in the case of wheat, the seeds sown and the crop harvested are very similar. In the case of a cow, the main activity of which is to breed calves, one approach to the production process could be to merely consider the cow as the means by which calves multiply. Hence, the self-reproductive process could be represented as a series of quantified, dated flows of homogeneous elements (Barceló and Sánchez 1988: 16-17). It may be, then, that in this similarity a source of confusion in the aforementioned quote of Georgescu-Roegen is found: the seed harvested is *not* the same as that which was sown, even though the latter is obviously grown from the former and both, within the bounds of genetic variability, may share the same chemical composition and physical appearance. Thus the seed sown is a pure input *flow*, incorporated into the output *flow* formed by the (other) harvested seeds. On the contrary, the lathe that enters *and* leaves the production process is the *same* machine, although it may have aged by one period.

The second part of the example given by Georgescu-Roegen is also of a certain consistency. In effect, alfalfa, for example, has an annual vegetative cycle although its great regenerative capacity permits it to re-grow if cut. Such cutting may be repeated once a month for some four years (with rest periods in those areas where the winter is harsher). However, if such cutting is stopped at the right time the plant will continue to grow and will bloom and produce seed. Thus the full life cycle of alfalfa offers two different kinds of output: forage produced by regular mowing and, optionally, a *final* growth cycle which produces seed. Thus the alfalfa seed may act as a *fund* which participates in generating two kinds of output *flow* which, moreover, are harvested at different moments in the *complete* production process. It is considered a fund for the same reasons as any other self-reproductive asset that operates as a fixed capital asset, be it laying hens, dairy cows or fruit trees. Thus, the position adopted by Georgescu-Roegen is not acceptable: clover seed produces forage repeatedly and thus acts as a fund. In each cycle (or period of time between the harvesting of forage) it enters and exits the process. Thus, given its nature as an active element over several periods, it cannot be classified as a flow.[9]

It should be noted that what is true of alfalfa or clover is also applicable to breeding animals when their meat is recovered when are finally slaughtered. Likewise, a fruit tree may produce its final output flow in the form of firewood, combined with its regular flows of fruit and firewood from pruning operations. In short, self-reproductive goods warrant a twofold approach:

1. In the case of single cycle organisms, the seed or animal used as an input shall be considered a flow, the output of which may consist of one or more products (as is the case with the wheat harvest which provides both grain and straw).
2. Those goods that have several productive cycles shall be considered funds, regardless of the number, moment or nature of the output flows they produce.

In any even, the analytical outlook of the funds and flows is to represent and study the (internal) organisation of production cycles (and the changes made thereto) in detail. As explained in greater detail below, the basic conceptual core of the model is any set of productive operations that permits a given output unit to be obtained. For this reason, the model is not

---

9 With everything stated so far, it is clear that Georgescu-Roegen was completely wrong with his example. Nevertheless, this mistake was probably an involuntary lapsus, the results of which have only survived as he never came to tackle the issue of self-reproductive goods in depth.

helpful for analysing the complex relationship between the economic and the biological life of self-reproductive goods.

1.2.3 Stocks and services

After classifying the elements involved in production, it is important to distinguish between the notions of funds and flows and the concepts of stock and services, respectively.

In the first place, funds are characterised by the enduring rendering of services. Consequently funds will neither grow nor reduce in size. In reality, they can only be put to better or worse use, and/or more or less continuously. For example, the full productive capacity of a machine will only be realised if it operates uninterruptedly at optimum power.

On the other hand, a stock is a *quantum* of substance, perfectly well located in space and time, for example, a granary, a water tank or a wine cask. Such stocks will grow (shrink) from (into) flows. Or, in other words, changes in stock levels derive from/cause flows. That is, stocks change by adding or subtracting the substance they consist of, at a rate (amount per unit of time) established as per convenience. Now,

1. Although all stock accumulates or diminishes in a flow, not all flows imply a stock increase or reduction. For example, electricity is a flow (kW/h) that does not come from a stock. It simply is generated by the transformation of the energy potential of a chute of water, or by using the heating potential of coal.
2. Any stock may accumulate or be reduced in *instants*, but not a fund: its use requires a specific time profile, in which there may be intervals of idleness. Georgescu-Roegen states that a bag of 20 sweets may make a similar number of children happy, either today, tomorrow, or distributed in any other way over time. Notwithstanding, a hotel room, with a useful life estimated at 1,000 days, will not be able to satisfy the pressing need of 1,000 tourists who have nowhere to spend the *coming* night. While the sweets constitute a stock that may be reduced completely in an instant, the hotel room is a fund, the use of which inevitably extends over a given interval of time. The same occurs with consumer durables: the only way to de-accumulate a pair of shoes is to walk in them until they fall apart (Georgescu-Roegen 1971).

A fund should not be considered a stock. To speak of a machine as stock makes no sense at all, unless it is withdrawn from production and held for spare parts. In reality, a machine is a service fund: a number of hours of activity distributed over a period of time in accordance with a certain pattern until such time as the machine is replaced for obsolescence and/or wear and tear.

Secondly, neither should there be any confusion between flows and the services provided by the funds. In effect, the two are accounted for differently. On the one hand, flows are expressed in terms of a number of physical units measured over a period of time. This rate of variance has a mixed dimension (substance)/(time). However, a slight problem does arise with respect to the length of the unit of time considered. If this is very long, the description of the flow may be too ambiguous. Thus, to speak of the construction of two oil tankers per year is an acceptable approach for a shipyard. To the contrary, to speak of a production rate of 10 million pencils per annum is too crude. In such a case the extent of the unit of time considered must be reduced (pencils per minute) and, to achieve a comprehensive description of the output flow generated, the pattern of changes in the rate of production over the year must be set out.

The magnitude of services rendered by the funds is calculated by the expression: (substance)x(time). There is no possibility of confusing them with flows. It is important to bear in mind the fact that the services, merely for conventional reasons, tend to be described by the expression (substance)x(time)x(time), i.e., referring to an extensive lapse of time. For example, a plant with 100 employees working 8 hours every day makes use of the services of 4000 man hours in a week of five working days. It is sufficient to multiply this by the suitable number of weeks to calculate the number of man hours available in a month (or year), a figure which tends to be used when making comparisons. In this point, unfortunately, malpractice does still occur when the number of working days or hours lost due to strikes, breakdowns, etc. is working out without specifying the number of workers affected. Such statements prevent one from obtaining a clear view of the magnitude of the variable under consideration.

With fixed funds, one also has to speak of a compound variable: machine-hours. Likewise, the rental of commercial premises is measured by surface area per month (or year). The services of rented durable assets are also paid for in terms of units-time. And, when not directly stated as such, it is simple to convert the terms given.

1.3 Temporal components of the production process

As with any process, the production process presents two attributes which reflect the determinant presence in it of the time dimension (Landesmann and Scazzieri 1996b: 192-3):

1. An initial moment and a final instant.
2. A given sequential order of phases. Any change to this arrangement would give rise to a different process.[10]

Given that all production processes are delimited sequences of phases of transformation, the time factor plays a twofold methodological role:

1. As a form of general chronological measurement. Defined as per the metrics of real numbers (Winston 1981: 13), time consists of a set of dates that permits events to be ordered sequentially, that is, ahead of, simultaneous to or after a given moment in time. In the model developed below, geophysical time units will have a twofold use (Piacentini 1996: 97): as an instant t of continuous time and as a span of time of greater or lesser length. The latter would correspond to an interval of time, such as a working day, month, week, year, and so on.
2. As a key variable in the description of the internal organisation of a production process. Time here represents an order that indicates changes of state. Or, in other words, as the duration that describes the development of a phenomenon.

A distinction should be made between the unit of time and the length of the period over which the average rate of change of the economic variables will be measured in a given analysis. Because this rate of change can vary, it is important to choose an appropriate unit of time and the length of the period. Let Q denote the output *per unit of time* of an industrial plant.[11] The production/month variable may be distributed over time in highly diverse ways, as shown in figure 3.

The product flow could be constant (figure *a*) over the whole span of time considered (for example, a month), or a flow which is concentrated over a few days (figure *b*) or a flow which varies throughout the period taken into account (figure *c*). Given that the specific pathway of the production variable within the defined time unit has implications with respect to the productivity and costs of the process, the choice of too long a time period could mean a loss of information and to erroneous conclusions being drawn with respect to the behaviour of the magnitude studied (Winston 1981). Notwithstanding, there is no time period which could be considered valid for all cases. Thus the only solution is to specify the unit used for each individual case: normally speaking, thus is an hour, working day or year. Given that time is a continuous variable, i.e., its hypothetical smaller

[10] Moreover, although there may be may loops, it is always feasible to find a sufficient lapse of time to ensure that the sequence of phases is not repeated.

[11] For instance, one month is the unit of time.

units exhaustively divide the larger, the use of these units of time should not present problems of coherence, although the greatest of care should be taken.

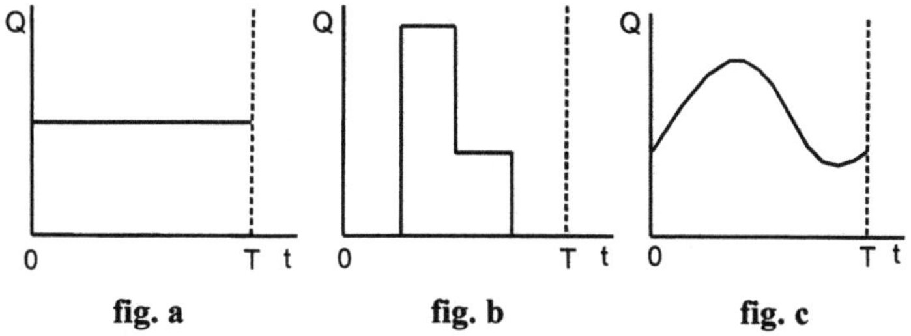

Fig. 3. The time profile of an economic variable

In this respect, with the exception of extremely long-term production processes (shipbuilding, aeronautical construction, large buildings, and so on), the rate of production tends to be expressed in terms of smaller units (hours, working days, weeks), while programmed production, or the time required to obtain a given output volume, is looked at in terms of longer spans (months, quarters or years). From an economic outlook, such longer intervals are also important as they reveal the influence of phenomena such as learning curves.

Due to a strange mix of convenience and a desire for exactitude, many microeconomic models consider production processes as instantaneous. Fortunately, cost analysis takes the time dimension somewhat more into account: short and long-term horizons are established. Although this is none too specific, it assumes that economic variables may change at least over the passage of time. However, the use of the logics of time as a heuristic procedure is not a perfect substitute for chronological time. In fact, economic phenomena always occur in real time and in order to perceive their changes fully it is necessary to use very different scales.

To sum up, using chronological time permits the order and duration of the different events involved in a production process to be established. Nonetheless, the span of the time unit against which events are referenced will vary according to the circumstances. Although there is a degree of freedom of choice, it is evident that the amount and quality of the information obtained will suffer dramatically if the time unit chosen is not appropriate. In this sense, experience, and the current convention, will indicate the best unit to apply.

Having set the initial (0) and final (T_p) instants of a production process, i.e., a given duration [0, T_p], it is generally possible to distinguish different phases or stages of this process. Such time intervals tend to be identified by the prominent role certain funds or flows may play therein. To obtain a better grasp of that internal structure, both in terms of time and substance, the concepts described up to now have to be broadened, thus providing a notably more detailed description of the production process.

The cornerstone to any technical and economic analysis of the production process is what is known as the *Elementary Process*. This is defined as *the deployment over time of the transforming activities in a given place which permits to obtain a unit of the product* (Scazzieri 1993: 86, 94-97). The manufacture of this separable output unit requires the start-up of a series of elements that constitute the minimum technical unit of the process (Morroni 1992: 23). Thus, on the one hand, a process on a smaller scale than would correspond to an elementary process is not economically justifiable (it lacks sense to produce half a car), albeit technically feasible. On the other, all the productive operations of an elementary process together configure a given state of the art.

In addition, it should be pointed out that an elementary process does not necessarily have to generate finished goods. That is merchandise with a presentation, composition and capacities that permit the consumer, or other firms, to buy it. Consequently, elementary process may refer to the production of parts, components, or pieces. The further processing of those goods will be continued by other establishments in the same *filière*.

Formally, a given production process could be defined as elementary if,

$$F_1(t)=0 \quad \text{for} \quad t\in[0, T) \text{ and } F_1(T)=1,$$

in which $F_1(t)$ is the output flow function (assuming *only* fully available at end) and T is the instant at which the process ends. In short, *elementary process refers to the set of productive operations required to obtain a unit of a specific good.*

In all elementary processes, the human work fund and means of production carry out highly diverse elementary or basic production operations. By means of these operations, and together with one or two flows, the final output is progressively created. Although not much attention has been paid to such operations herein, it must be acknowledged that elementary operations have been the subject of the most detailed examination in time and motion studies. For the purpose of the present analysis, it is sufficient to establish the criterion that a production operation shall be considered elementary when any attempt to further break it down merely succeeds in creating unnecessary additional work. Or, in another words, an elementary operation is defined as the smallest possible unit of work.

For practical reasons it is pertinent to group all such simple operations in tasks together. A task is defined as a completed operation usually performed without interruption on some particular object (Scazzieri 1993: 84).

Or,

Tasks (...) have to be defined both in relation to the stages in the processes of transformation of specific materials as well as in relation to the activities of fund-input elements. The specific skills and capacities of different fund elements constrain the specification of tasks and the type of task arrangement that can be implemented in a productive process (Landesmann and Scazzieri 1996a: 195).

To define tasks a verb is needed to denote an action, together with the precise identification of the object and the way in which it is affected.

Despite the definition, the identification of tasks is, to a certain extent, arbitrary. A practical criterion is to delimit them in the same way as those who usually perform them. Fortunately, in many production processes, a job tends to be associated with a particular task, making it easier to identify. In any case, it should not be forgotten that the specific content of tasks, or set of elementary production operations, is modified according to the degree of the technical division of labour, in line with the technical changes that occur (Gaffard 1990: 102). Consequently, although analytically speaking it may be opportune to consider tasks as the minimum units of a production process, it would be foolhardy to expect that over time the type and number of basic operations would remain unchanged. Any technical innovation, to a greater or lesser extent, will lead to a redefinition of tasks.

The instruments used in a task do not define it. In effect, the same task may be performed with different tools according to the technical level of the process and the skill of the worker. Nonetheless, the same tool may be used in a broad variety of different tasks. Such is the case of tools that have a broad range of use, like hammers or knives. There is therefore no one-to-one correspondence between tasks and tools.

With respect to fixed funds, the question of the level of activity has two dimensions (Scazzieri 1983: 600, 1993: 85-86):

1. A machine is fully occupied if, at any given moment, it intervenes in the maximum number of elementary processes it is designed to act on simultaneously. Such a level of activity, considered *normal,* may be measured in one or more of the appropriate technical units, such as revolutions per minute or volume of current product processed.
2. A machine is considered *continuously* employed if it is active throughout all the time available.

For the case of a single machine, see figure 4, which mixes, in simple terms, the dimensions of the intensity and time of activity. The striped area shows use below full capacity (p_c), the magnitude of which is measured along the x-axis. Meanwhile, the lapse of time between t_1 and t_2, placed within the $[0, J]$ interval (that is, the working day), indicates the idle of the unit in question. It should be pointed out that, from the point of view of the production process as a *whole*, the use of fixed funds has a third dimension: the operationality of the units available.

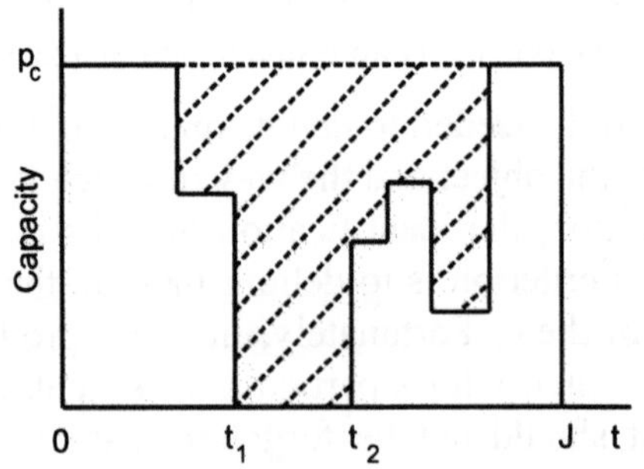

Fig. 4. The two dimensions of fixed fund activity

Paying attention to figure 4, the conclusion is reached that the full use of a capital asset implied continuous activity. Notwithstanding this, the opposite would not be true. In effect, an interruption in operation would represent a level of non-use while a greater or lesser excess capacity is compatible with permanent operation.

Tasks adjacent in time and space may be grouped in phases (or stages). The perspective of the engineer is of great help for recognising such phases. In effect, technical manuals on production processes normally tend to consider the different forms of treatment applied to a specific item as stages or phases. For instance, chemical processes always include such basic operations as pulverising, mixing, heating, filtering, precipitation, crystallisation, dissolution, and so on, which are clearly differentiated, both by their order in time and for requiring the use of specific equipment.

In the case of products arising from getting-together of highly diverse subsystems, components and/or parts, neighbouring tasks along the process may be grouped in phases. As is logical, these different phases may be performed by one single firm in one facility or by many different companies with plants located worldwide. In addition to spatial separation, the different interim processes may also be distributed over time. The case of the automobile industry may be considered as an example of several stages in a complex assembling process. A minimum of six distinguishable phases could be taken into account: the machining of the engine block and the fitting of the basic engine components, the stamping of the bodywork with presses and dyes, the assembly of the bodywork (mostly with robot welding, though some operations are still manual), painting (with several coats of different kinds of product), the mounting of various other units (engines, axles and transmission systems) and, lastly, the manual assembly of the interior (seats, dashboards), and final inspection.

In short, the identification of *the phases coincides with the sequence of the major operations of transformation which the product in process is subjected to*. This may be represented in diagrams of diverse complexity. On occasions, such diagrams will be basically lineal, as is the case of the manufacture of concrete while, on others, they will be basically tree-like, as the different phases converge towards the final assembly. Such is the case of home electric appliances or electronic devices. Or they may diverge from an initial raw material, as is the case of the derivatives obtained from the successive cracking of crude oil (Romagnoli 1996b: 11-12).

An additional consideration is the fact that in numerous production processes the phases are separated by stocks of goods in process. The holding of such stocks may generally be justified for technical (for example, waiting for adhesives to harden or paint to dry) or organisational reasons (to avoid bottlenecks, resolve time lags and so on).

The phases, considered as sets of adjacent tasks, may be recomposed. That is, the order and number of tasks contained therein may be altered (Piacentini 1995; Landesmann and Scazzieri 1996a).

The last term to define is *Production Process*. When somebody speaks about a production process, what it is generally understood is the current activity of a farm or an industrial facility. Production process *is the stream of successive elementary processes carried out by a productive unit or plant* that gives a definite volume of output per unit of time, or, in other words, the number of items of one or several ranges that are periodically generated by a given establishment. As is known, this is something that invites one to think about the technology used, the number of employees, the level of cost, etc. In short, the concept of a production process refers to the most directly observable aspect of productive activity: a running succession of elementary processes which could be deployed in different forms. According to Georgescu-Roegen (1969, 1971), such arrangement may be of three different kinds: sequential, parallel or in line. These issues will be dealt with in chapter 3.

The conceptual chassis established above comprises the following elements:

Elementary operations $\subseteq$ tasks $\subseteq$ phases $\subseteq$ elementary process $\subseteq$ production process

The analytical value of such segmentation of production activity resides in stressing its complex internal nature. To the contrary, the only use of simple chronological units (hours, minutes or seconds) to describe a production process suggests a misleading homogeneity.

1.4 The table of productive elements

All elementary processes can be represented by tables. Each element in the table will indicate the input/output rate of a flow, or the interval of time for which a fund is in use for each task (or phase). This is a very simple concept, which, despite not being explicitly suggested by Georgescu-Roegen, was latently present in his work (Morroni 1992: 75). Thus, the idea is to use the bi-dimensionality of the table to indicate the magnitude of the flows and/or funds present in the different stages (phases or tasks, as applicable) of an elementary process. As is clear then, the table informs us of the basic technical and economic features of the production process. Any change to the material used, the operating times of the different machinery, or whatever, will require a new table. The general form of such tables is shown in the next page.[12]

The table comprises two parts. The top is devoted to the times the funds are activated in the phases (or tasks) of the elementary process. Denoted as $N_{j,s}$, elements ($n_{j,s}$) indicate the time for which the fund j is in service ($j=1, ..., J$) in phase s ($s=1, ..., S$). Thus, for example, in each and every one of the stages in which the services of a given machine are required, the number of hours of use is indicated. The empty boxes will show the stages when that particular fund is not in use. The human labour fund will most probably appear in all stages, although with differing values. In turn the plant (measured in hours of activity) is always present and always with the same value. Finally, one or more of the rows of the sub-table will refer to the times which the different stores are in use for.

The lower part corresponds to the sub-table of flows per phase (or task). It is named $A_{k,s}$, and contains the elements ($a_{k,s}$) that show the amount of flow k ($k=1, ..., K$) fed into, or produced during the stage s ($s=1, ..., S$). Given that the chronological order of the phases is to be maintained, the *outflows* appear in the final column of the sub-table (with rows filled with zeros when referred to the previous stages of the process). These outputs include the main output, by-products and residues, waste and emissions. In

12 The table of the different elements of staged production is directly based on the work of Piacentini (1995) and Morroni (1992: chap. 7), whose findings were very similar. Nevertheless, both authors use the term "matrix" instead of "table" which is preferred here. The reason is that the concept is merely descriptive, and lacks the qualities required for mathematical manipulation.

case of different outflows produced in the interim stages of the process, some rows would be added to indicate each of them.

$$\begin{bmatrix} n_{11} & n_{12} & \cdots & \cdots & n_{1s} \\ n_{21} & n_{22} & \cdots & \cdots & n_{2s} \\ . & . & . & & . \\ . & . & & . & . \\ n_{j1} & n_{j2} & \cdots & \cdots & n_{js} \\ a_{11} & a_{12} & \cdots & \cdots & a_{1s} \\ a_{12} & a_{22} & \cdots & \cdots & a_{2s} \\ . & . & . & & . \\ . & . & & . & . \\ a_{k1} & a_{k2} & \cdots & \cdots & a_{ks} \end{bmatrix}$$

Sub-tables $\boldsymbol{N}_{j,s}$ and $\boldsymbol{A}_{k,s}$ may be grouped together as a single table $\boldsymbol{P}_{j+k,s}$, given that the number of columns coincides. That is,

$$\boldsymbol{P}_{j+k,s} = \begin{bmatrix} \boldsymbol{N}_{j,s} \\ \boldsymbol{A}_{k,s} \end{bmatrix}$$

$\boldsymbol{P}_{j+k,s}$ is the *table of the elementary process*. The rows consecutively show the funds and flows required for the successive phases of the process, which have been shown in columns. It should be noted that the values read per row, corresponding to the inputs employed by the process, may be added together. The sum then gives a vector which indicates the physical amount of inflows and the service times of the funds employed in the elementary process. Although this information allows a detailed knowledge of the process, it is not sufficient to give a very detailed insight into its internal organisation. In order to achieve a complete description, the model proposed has been build considering the two dimensions shared by all production processes: an account of different elements and the presence of time.

The sum of columns is impossible, given the heterogeneity of the elements shown. Notwithstanding, this limitation has its own kind of "silver lining". Indeed, the simultaneity of diverse funds and flows in each stage of the elementary process underscores an essential feature of all productive activity: the fact that the different elements *co-operate* with each other. That is, the complementarity of the factors of production is an earlier, and more basic, quality than any possible substitutive relationship that could occur. It is sufficient to read the table by its columns to realise the essential reciprocity of the elements of production, given that the collaboration between them is patent through each and every one of the different phases of the production process. For this the table allows us to perceive the fabric woven by the funds and flows. This is a fundamental feature, which is hard to observe by means of the simple side-by-side layout proposed by the production function model.

As stated above, it is assumed that the confines of the phases are established according to the analytical goals of each individual research project. At all events, it is best not to multiply the number of phases, as this would diminish the main quality of the elementary process table: to see the complete organisation of a production process at a glance (Piacentini 1995: 471).

Despite being arranged as per phases (or tasks), the table does not explicitly show real time intervals. Therefore, it should be supplemented with an auxiliary chart containing the detailed production operation times. This way, it is possible to indicate the precise chronology of the activation of the funds, given that duration of the service rendered thereby could be less than that of the stage in which they operate, and the precise instant at which the inflows are incorporated. To all events, the greater the need for detail, the harder it is to understand and manipulate the information in the elementary process table.

The table may equally refer to an *ex-ante* as an *ex-post* elementary process. The former would be the case of a general organisational design for a future production activity, while the latter would recount the use of funds and the consumption of flows per output unit produced by an existing process. In either case the different tables generated would be comparable.

The elementary process table is not only applicable to manufacturing processes but may also be applied to the rendering of many different types of service, such as transport, hotel and catering, although certain variations may need to be incorporated to aid a more global understanding of the phenomenon.

Here it should be remembered that the elementary process table enhances the information supplied by the technical coefficients. As is known, technical coefficients are calculated from magnitudes expressed in terms of value. This could be misleading. But the main shortcoming is not to consider their temporal dimension. This may be a problem if the aim is to do an in-depth study of technical change. A simple way to obtain a comprehensive definition of the technical coefficients (a_{ij}) of an input-output table would be to consider $\mu_{ij}>0$, i.e., the length of time between the moment the inputs enter the process and the instant the output is obtained, as the unit of time (Lager 1997: 359). So, at moment $(\tau+\mu_{ij})$, an output volume of

$q_j(\tau+\mu_{ij})$ units will have required a total of $a_{ij}q_j(\tau+\mu_{ij})$ input units. Thus, calling this amount $v_{ij}(\tau)$, we obtain the input requirements at moment τ.

As the technical coefficients tend to be calculated on the basis of yearly data, the total volume of units of merchandise *j* produced in a year is given by:

$$q_{j,t} = \int_t^{t+1} q_j(\tau)d\tau$$

while the operating inputs used in the same period are,

$$v_{ij,t} = \int_t^{t+1} v_{ij}(\tau)d\tau = a_{ij} \int_t^{t+1} q_j(\tau + \mu_{ij})d\tau$$

Thus, the new technical coefficients are obtained by comparing the yearly production with the annual consumption of inputs. That is:

$$a_{ij}^* = \frac{v_{ij,t}}{q_{j,t}} = a_{ij} \frac{\int_t^{t+1} q_j(\tau + \mu_{ij})d\tau}{\int_t^{t+1} q_j(\tau)d\tau}$$

This way, the coefficients reflect both the technical requirements of production and the time profile of the process. However, the above device is based on the non-realistic assumption that the inputs (operating) enter simultaneously into the process. Nor does it consider the activity times of the fixed inputs and labour, both of which are susceptible to modification through technical or organisational change.

Meanwhile, returning to the table of productive elements, one sees at a glance that it contains more technical/economic information than the mere technical coefficients. For example, the funds sub-table may provide detailed information on the degree of funds specialisation. In effect, if the degree of specialisation is at its highest, all the components not placed on the main diagonal line will be nil. Moreover, if certain funds are inactive during part of the duration of one or more phases, problems are revealed with respect to the efficiency of the process. This is similar to what occurs when comparing two tables relating to the production of a given quantity of the same goods: greater inefficiency could be associated with a higher consumption of flows.

Prior to winding up this section, it is important to stress the following: the elementary process table is *not* of great interest from an engineering point of view. Indeed, the technical behaviour of the funds is far from sufficiently clearly described by the mere times of activity shown in the table.

Likewise, the entry rates of the flows do very little to describe their exact function within the process. In the case of the output flow, the amount obtained tells us nothing of its design or functionality. A detailed description of the capacities of a fund, and an assessment of its reliability, require a battery of technical measurements far in excess of the information given by the table. With respect to flows, there are no details referring to their physical nature either. A glance at any engineering manual would reveal just how different it is to the funds and flows model. It would be a crass error to declare that the analytical theory of Georgescu-Roegen enters into a terrain alien to economic analysis. Despite the model being far from a strictly engineering outlook, it offers a more realistic approach to the production process. Hence, an engineer will probably pay greater attention to it than to others of a very abstract nature.

1.4.1 Applied experiences

The model presented could easily be adapted to the applied analysis of the production process. In this sense, various authors have suggested the suitability of the Georgescu-Roegen approach and later developments of this, for the representation of the production process and any possible technical modifications thereto (Piacentini 1987; Gaffard 1994). But others have gone far beyond this assertion, and have created and used adapted versions of the funds and flows model to analyse the efficiency of such diverse processes as farming (Romagnoli 1996a; Zuppiroli 1990), telecommunications networks (Marini and Pannone 1998), or industrial processes (Birolo 2001; Mir and González 2003: 57-64). However, it is to the work of Morroni (1992, 1996, 1999) that greatest attention should be paid to. It includes a computer programme called *Kronos Production Analyser*™ (Moriggia and Morroni 1993; Morroni and Moriggia 1995), designed to systemise information relating to the basic technical and economic features of production processes. This packet is fed with data gathered by way of a standardised questionnaire. The software is then used to make whatever calculations the research may require. The output obtained consists of four tables showing the most significant features of the process studied in a clear, ordered and exhaustive way. There is no doubt that such tool and its outcomes facilitate a better understanding of production processes and may improve the study of process innovations, i.e., the changes in the capabilities of the different elements of production (Loasby 1995).

The basic idea behind the *Kronos Production Analyser*™ is to build a table similar to the $\boldsymbol{P}_{j+k,s}$ shown above. From this technical table, another may be derived with economic measurements in which all of the components are multiplied by their respective market or attributed prices (Morroni 1992: 82). Such prices are, of course, merely exogenous data. Nonetheless, these conceptual tables need to be adapted to be used in applied research. This implies the development of an extensive battery of technical-economic concepts related to the most salient variables of any production process. The following paragraphs are devoted to this.

Firstly, it is established that the salary costs plus the cost of the input flows make up the *Direct Costs.* Adding the latter to the cost of maintenance and depreciation of the machinery, the *Transformation Costs* are obtained. This amount, when it is added to the costs of storage, plus others related to real estate, provides us with the *Industrial Costs.* The margins have been calculated with respect to the prices before commercial distribution. Thus, the difference between the latter and the industrial costs gives us the *margin on industrial costs.* The other pertinent margins are the *margin on transformation costs* and the *margin on direct costs.* It should be pointed out that all the above are gross margins as, given the nature of the research, no information is gathered with respect to the firm's interest expense or fiscal burdens. The aim is not to emulate company accounting, but to obtain a detailed picture of the technical and economic situation of a specific process.

Another point to bear in mind is the time during which the equipment is in operation. With respect to the activity of fixed funds, the basic interval defined is the *net machine time,* or the time effectively invested in process operations by a given machine. On either side of the above interval, we have the loading and unloading times. The sum of the three constitutes the time required to produce one output unit. If the lead time, that is, the time required for the machine to be readied for service, or to be adjusted for a change of batch, is added to this, we obtain the *total machine time.* Finally, adding the above times to the maintenance time will give us the *gross machine use time.*

Information is also gathered with respect to the flexibility of the process and the time intervals. The latter, given their importance, are listed below in simplified form and are also shown in figure 5 (Moriggia and Morroni 1993; Morroni 1992: 72-75, 1996):

1. *Net (Gross) Process Time*: the period that an elementary technical unit requires to produce an output unit, excluding all interruptions for loading and unloading machines, lead time, maintenance, and so on. In short, this is the sum of the intervals of physical and/or chemical processing of the product in process. Gross time will include the technical storage of the inputs or outputs.

2. *Working Time*: the above periods plus the storage of semi-manufactured products required for breaks of continuity between the phases of a process (or organisational stock).
3. *Net (Gross) Process Duration*: in net terms, this includes the working time plus the initial warehousing of inputs and final storing of finished plant products. *Kronos Production Analyser*™ considers the longest interval under normal conditions. In gross terms, process duration includes the response time added to net duration.
4. *Response Time*: This is the lapse in time between the reception of the order and delivery to the client.[13]
5. *Duration*: total time from the first input entering the process to obtaining a totally finished product. This includes all kinds of interruptions, including those caused by the need to adjust the equipment for changes in output-mix.

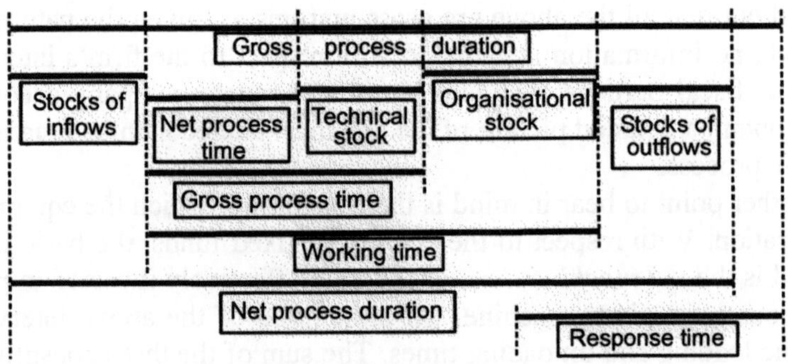

Fig. 5. Process times

In addition, it should be pointed out that some of the time intervals defined above may coincide in a given process.

On the other hand, *Kronos Production Analyser*™ distinguishes two main types of stock: technical (the product in process needs to be stored to mature, for decanting, and so on) and organisational (stored separately from the different input flows, such stock corresponds to the stock of finished product and others resulting from bottlenecks between the different phases of the process).

As stated above, there are two questionnaires to systemise the collection of data in the field, which must be entered into the programme. *Kronos*

[13] Other times are *Gross Duration,* which comprises the time between the inputs being ordered and received and the *Financial Time,* or mean lapse of time between paying for the input and being paid for sales.

Production Analyser™ first generates the *Quantitative and Temporal Matrix* (Morroni 1992: 86-92). See table 1.

The amounts shown in the table are those required to produce a unit of the good. The rows show the quantities of the elements of production required, while the columns detail the different phases of the production process.

Table 1. Quantitative and Temporal Matrix

Description of the element	Quantity	Price/cost
	a) Outflows	
	b) Inflows	
	c) Fund Services	
	c.1) Workers	
	c.2) Equipment	
	c.3) Inventories	
	c.4) Plant	

Moreover, another table is constructed containing additional information on the features of the labour force, the administrative regulations, or the market in which the plant under study operates. This is the *Organisational Scheme* (Morroni 1992: 92-98). The table is divided into two blocks:

1. The first shows the pattern of use of the services of human work. For this, it shows the distribution of the workers in terms of category and working times (working day and shift). It also contains additional information, such as age, sex and educational distribution. Finally it shows the general distribution of tasks for each category of worker.
2. The second block refers to the machinery used, listed by type. For each type, the lead, loading, process, unloading and maintenance times are described, along with the operation speed.

In short, an *Organisational Scheme* of a process would appear as indicated in table 2.

Table 2. Organisational Scheme

a) Workers
a.1) Number of workers by workplaces, shifts, etc.
a.2) Tasks and skills by workplaces
b) Equipment (machines)
b.1) Number by type
b.2) Time of use
b.3) Intensity of use

The *Quantitative and Temporal Matrix* and the *Organisational Scheme* give a good snapshot of the production process.

As has been said, it should be noted that, in addition to its capabilities as an accounting instrument, *Kronos Production Analyser*™ also offers a powerful structure with which to tackle a series of questions regarding process innovation at a microeconomic level, for two reasons:

1. First of all its descriptive capabilities. In effect, the model is well suited with respect to the internal organisation of the process and it also explicitly features the time factor, expressed as duration. It includes the technical features of the elements involved, those of the product, and an economic valuation of the process.
2. It also permits empirical research on a plant (enterprise) level, i.e., at the level of the locus at which innovation is decided upon and implemented.

Kronos Production Analyser™ provides an improved approach to the reasons for, and consequences of, process innovation, since it offers a highly detailed dissection of production activity. It is thus clear that the programme may be deployed in either of the two following ways (Loasby 1995):

1. Transversely: simultaneously researching different production units working on the same process. This application is of indisputable interest for comparative studies of individual plant efficiency, analysing economies of scale, assessing the pros and cons of the indivisibility of certain factors, and so on.
2. Longitudinally: studying the modifications of a given process over time. To such ends, this line of research would select a single plant and study it over the years.

In either case, although the nature of the information gathered means that the fieldwork is not too complicated, especially if the firm is experienced in cost accounting, it must be acknowledged that the data required tend to be confidential and company management may thus be reluctant to provide access to it.

1.5 Representation of the elements of the production process

Georgescu-Roegen never put forward the elementary process table. He chose an undoubtedly more visual, but more laborious method to represent the activity of funds and flows. In effect, in his representation, the participating elements were considered *at each and every instant of the duration of the elementary process.* To these ends, their activities were shown through functions, the graphic representation of which is quite singular. Specifically (Tani 1986: 200):

1. Taking the kth element ($k=1, ..., K$), the function $I_k(t)$ indicates the amount of the element that enters the process between 0 and t, for each instant $t \in [0, T]$.
2. Taking the kth element ($k=1, ..., K$), the function $O_k(t)$ shows the amount of the element that exits the process between the initial instant and t, where $t \in [0, T]$.

The functions $I_k(t)$ and $O_k(t)$ are accumulative and thus do not decrease in the $[0, T]$ interval in which they are defined. It may be assumed that $I_k(0)=O_k(0)=0$. These functions do not tend to be continuous given that a finite quantity of an element enters or exits in a single instant.[14] So a production process may be represented by two vectors of time functions:

$$\{I_1(t), I_2(t), ..., I_k(t)\} \; ; \; \{O_1(t), O_2(t), ..., O_k(t)\}$$

or,

$$\{I(t); O(t)\}, \text{ where } t \in [0, T].$$

This representation is overabundant since it contains several functions which are identically nil. This is the case of flows, which either only appear on the input side (to be processed), or on the output side (emerging from the process). On the other hand, funds are both inputs and outputs, and comply with the following:

$$I_k(t) \geq O_k(t) \text{ and } I_k(T) = O_k(T), \text{ where } t \in [0, T].$$

The awkwardness of the graphic representation suggested led Georgescu-Roegen to propose a more manageable form of visualisation. To start with, the flows, measured in their respective physical units, are denoted by the accumulative functions $F(t)$. There are two cases (Tani 1986: 202; Morroni 1992: 55):

1. $O_k(t) = F_h(t)$ ($h=1, ..., H$) when it is purely an output flow (or *outflow*).
2. $I_k(t) = -F_i(t)$ ($i=1, ..., I$) if the flow is only an input flow (or *inflow*).

[14] If this occurs in the vicinity of the initial instant, then $\lim_{t \to 0+} I_k(t) > 0$, with the function showing a point of discontinuity at origin.

$F(t)$ functions are monotonic and not decreasing given that, in order to simplify their graphic representation, the values are considered in absolute terms. An example of a $F_i(t)$ function is shown in figure 6.

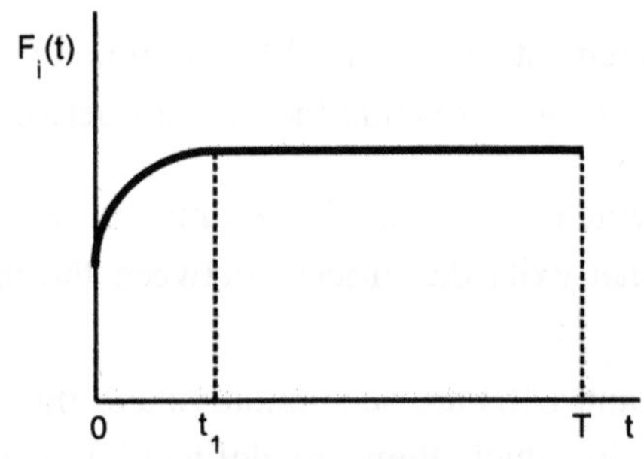

Fig. 6. The $F_i(t)$ function

In the $[0, t_1]$, successive amounts of a given inflow gradually enter the process. On the contrary, in period $[t_1, T]$ the function becomes flat because no more units of the element are incorporated into the process. Changes in the amounts and times of the absorption of flows as well as outflows will alter the pathway of the functions $F(t)$.

With respect to funds, given that these are characterised by the fact they enter and exit the process, functions $U_j(t)$ (j=1, ..., J) are defined as follows:

$$U_j(t) \equiv I_j(t) - O_j(t),\ t \in [0, T]$$

As a general rule, function $U_j(t)$ is equal to 0 when the jth fund is not active. If it is in operation, the value of this function will be a positive number equal to the number of units involved in the process. This means that $U_j(t) \geq 0$ for $t \in [0, T]$ and, moreover, given the prior definition, $U_j(0) = U_j(T) = 0$, i.e., in the initial instant it has not yet entered the process while, in the final instant, it will already have exited it. However, the case may also be that,

$$\lim_{t \to 0+} U_j(t) > 0 \text{ and } \lim_{t \to T-} U_j(t) > 0.$$

A hypothetical example of function $U_j(t)$ appears in figure 7. It is a hypothetical fund that is active for interval $(0, t_1]$ while for interval $[t_1, T]$, it is not participating in the process.[15]

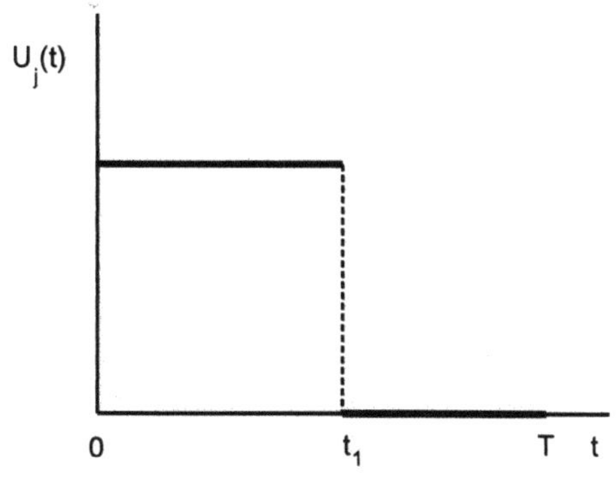

Fig. 7. The *Uj*(*t*) function

One final way of representing the services of the funds is by function $S_j(t)$ ($j = 1, ..., J$) defined as follows (Tani 1986: 205-206):

$$S_j(t) = \int_0^t U_j(\tau) d\tau$$

$S_j(t)$ measures the accumulated total of services rendered by a given fund between 0 and t. Or, in other words, the time the fund has been active in the process from the beginning through to the moment at which the measurement is taken. Function $S_j(t)$ is not diminishing and is always continuous. Moreover, $S(0) = 0$. In the intervals in which the function $U_j(t)$ is zero, the function $S_j(t)$ is horizontal because no services are accumulated. On the other hand, in the intervals in which the function $S_j(t)$ increases, the fund is active. Point T will always form part of the maximum values of $S_j(t)$. This is shown in figure 8.

Having reached this point, the next step is to represent a production process by a flow functions vector and a fund functions vector. That is, a *functional*: the relationship between the main output function and the *set* of functions that represent the remaining funds and flows, that is,

[15] The measurement of the services of the funds by way of functions $U_j(t)$ offers the benefit that any change to the number of units employed and/or the frequency of activation can easily be shown.

$$\overset{T}{\underset{0}{Q}}(t) = \Psi\left[\overset{T}{\underset{0}{F}}(t); \overset{T}{\underset{0}{U}}(t)\right]$$

This functional represents the

catalogue of all feasible and non-wasteful recipes (Georgescu-Roegen 1971: 236, underlined in the original).

That is, the complete set of production processes from which one will be chosen according to certain criteria.[16]

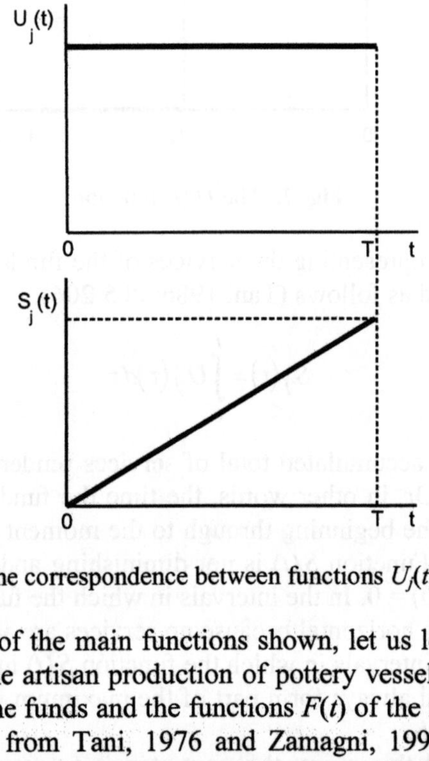

Fig. 8. The correspondence between functions $U_j(t)$ and $S_j(t)$

As an illustration of the main functions shown, let us look at the elementary process for the artisan production of pottery vessels. In this case, the functions $S(t)$ of the funds and the functions $F(t)$ of the flows are shown in figure 9 (adapted from Tani, 1976 and Zamagni, 1993). The horizontal axis of the figures represents time while the vertical axis indicates the amount accumulated (physical or services) of the element in question. This amount is measured in technical units (Kg., kW/h., m^2, etc.) although, for funds, it seems more appropriate to indicate the number of units present. The process in question has the following six phases: the first operation

[16] This functional corresponds to the traditional production function when time is introduced according to the fund-flow approach. Moreover, this functional could be associated with a *set* of the previous defined table of productive elements ($\boldsymbol{P}_{j+k,s}$).

consists of the worker entering the scene and taking the clay paste and water required for the process. The task occupies the interval $[0, t_1]$. Then the wheel is used for the period $[t_1, t_2]$ to shape the clay. At the end of this operation, the potter leaves the vessel to stand on a shelf momentarily $[t_2, t_3]$. Then the kiln is fired up and operates during $[t_3, t_4]$, consuming electricity. Once the piece has been fired, the worker enamels it, a task which occupies interval $[t_4, t_5]$. The kiln is again used for a second firing, the duration of which is $[t_5, T]$. For the sake of simplicity, the figure does not contemplate the operations of loading and unloading the shelves and kiln. At the end, instant T, the finished product is obtained, a pottery vessel. The whole process, $[0, T]$, takes place within a given space, the workshop, of given dimensions, the presence of which should also be taken into account.

1.6 General properties of the elementary process

Divisibility, homogeneity for integer numbers and fragmentability are three of the general properties of production processes, which, due to their relevance, deserve to be tackled individually. In addition to defining the terms, this section shows their forms of representation and analytical implications.

1.6.1 Divisibility

An elementary process represented by functions,

$$\{F(t),\ U(t)\}$$

is considered divisible if there is one or more positive integer numbers $n>1$ so that the resulting process,

$$\left\{\frac{1}{n}F(t), \frac{1}{n}U(t)\right\}$$

is viable. It may be activated in fractions, reducing the amounts for all elements proportionally. It should be noted that the property of divisibility does not require the process to be so for any $n>1$. Obviously, the broader the range of such numbers is, the greater the degree of divisibility of the process. To all events, this condition is more relaxed than it usually is in conventional production models. In effect, in the latter, a process Z is only divisible if λz is a feasible process, where λ is any real number from 0 to 1.

In the case of the funds and flows model, the condition is less stringent as the assumption of perfectly divisible inputs has been dispensed with.

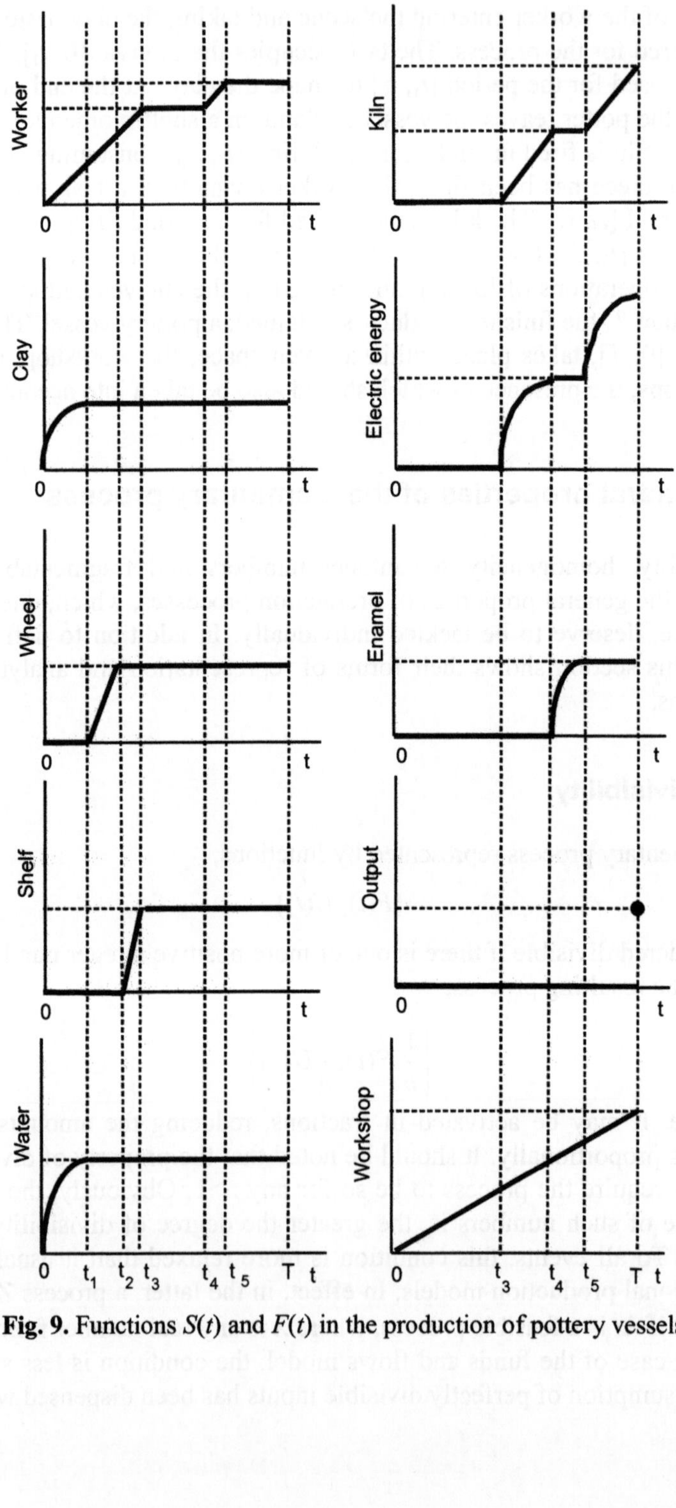

Fig. 9. Functions *S*(*t*) and *F*(*t*) in the production of pottery vessels

1.6.2 Homogeneity

Elementary processes have the property of being homogeneous for integer numbers, that is, given a process,

$$\{F(t), U(t)\},$$

will always be a viable process like,

$$\{n \cdot F(t), n \cdot U(t)\},$$

n being a *positive* integer number. This property has been called "constant yield for integer numbers" (Tani 1986: 212). It must be pointed out that although a process may be feasible, it is not necessarily efficient.

1.6.3 Fragmentability

An elementary process $\{F(t), U(t)\}$ is fragmentable if there are *H* possible sub-processes denoted as,

$$\{F^h(t), U^h(t)\}\ (h=1, ..., H>1),$$

and not all necessarily of the same duration (Tani 1976: 131, 1986: 211), so that,

$$\sum_h F^h(t) \equiv F(t) \text{ and } \sum_h U^h(t) \equiv U(t), \text{ for all } t \in [0, T].$$

Fragmentability allows us to identify the different stages of an elementary process. Nonetheless, what may be stressed is the fact that each phase may be activated separately. This means that the phases must be of sufficient technical entity to be able to operate alone. This capacity will depend on the supply of interim goods from the preceding phases. This is a property that permits the phases to be activated at different times and places (establishments), even when a rigid consecutive order is established for technical reasons.[17]

Fragmentability is the attribute which means that many productive processes can be carried out in different plants worldwide. It guarantees obtaining certain advantages that result in reduced production costs, despite also causing problems with respect to co-ordination and logistics. To all events, it must be remembered that not all production process are equally fragmentable, although the circumstances may change over time.

[17] By definition, this property is not attributable to the tasks. The latter form a compact block in the phases they are contained in.

Obviously, in a fragmented process, the output of one phase is the input of the next, except for the last, the output from which is the finished product. This then requires the introduction of a new element of the production process: interim outflows, the properties of which are:

$$\sum_h F_s^h(t) \equiv 0 \equiv F(t) \text{ for } 0=t=T,$$

given that they are outflows (denoted as positive) of one phase and inflows (shown as negative) of the following.

On occasions, such interim flows will result in a need to hold an interim stock, which could be considered as another phase in the process. Such a need appears when there is a time gap between the instant at which the product in process leaves one phase of activity and the moment when it enters the following stage of the process as an input. As has been stated, this can basically be justified by one of two reasons:

1. Because of the product in process needs to be decanted, settle, ripen, etc.
2. Because of the phases are out of step.

Apart from the stock held between phases, there are others which comprise the raw material prior to processing and stocks of finished products, ready to be sold within a reasonably short period of time. Both may be considered as phases of an elementary process.

Formally, a unit of the product in process which constitutes an elementary process, will remain in an interim store for a given interval of time,

$$[T_0^a, T^a] \subset [0, T]$$

Thus the storage time for inflows and outputs is likewise considered in the duration of the elementary process.

To analyse the behaviour of the storage phase, it is enough to observe the inflow and outflow of goods from a warehouse over a day or any other standard time unit. Therefore, the dynamics of a stock during the course of a working day $[0, J]$, and for a flow F_i of product in process, could be represented as in figure 10 (Tani 1986: 213).

The figure has been constructed under the assumption that the store operates according to the *first in, first out* (fifo) rule, and that each day the store fills with, and empties of, the units that enter on that day. In this case, the term F_i^1 represents the accumulative function of the output from the previous phase. That is, it shows the increase of the product in process stock, fed by the inflow. On the other hand, the term $-F_i^2$ represents the accumulated input function addressed to the following stage, or, in other words, the product entering the following stage or, what amounts to the same, the interim outflow leaving the store. Following the convention, this outflow is represented by a negative number. The points on the *x*-axis t_2 and t_1 indicate the end of the output of the previous phase and the start of the input of the following phase respectively.

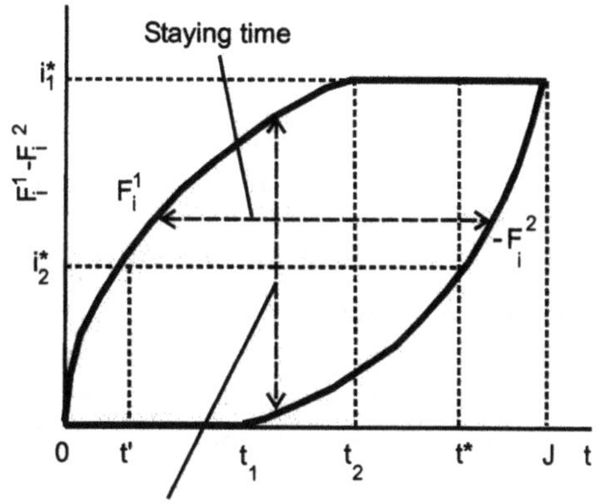

Fig. 10. The storage phase

With the *x*-axis giving the time and the *y*-axis showing the amount of product, a reading of the horizontal distance between the points of functions F_i^1 and $-F_i^2$ gives us the time the product in process is held in stock, while the vertical distance between the functions gives a measurement of stock at any given moment in time. For example, the distance between *t'* and *t** shows the time for which amount i^*_2 remains in stock, while at *t** the difference i^*_1-i^*_2 indicates the amount of an element held in stock. In short, the vertical distance provides an indication of stock levels while the horizontal provides the speed at which the stock is emptied.

On the one hand, by calculating the mean of the horizontal distances, the average length of time for which the product remains in the stores (π) is obtained. On the other, the mean of the vertical distances gives the average volume of stock held (ϖ). Given that π is equal to the surface area contained between the two curves divided by $F_i^1(J) = F_i^2(J) = F_i(J)$, while ϖ is equal to the same area divided by J, then,

$$T \cdot \pi = F_i(J) \cdot \varpi$$

The above expression brings together the duration of the storage phase, the average time in stock, the total inflow and outflow for a day, and the average stock level.

1.7 Limitations of the model

When attempting to model a production process, one possible option is to consider all participant elements as flows, while another is to combine funds and flows. The former is the option chosen by such diametrically opposed rivals as, on the one hand, the von Neumann and Sraffa models and, on the other, the production function and the Koopmans's activity analysis. The strongest defendant of the latter is Georgescu-Roegen, although there is an initial version of his model formally expressed only in terms of flows: the case of the above defined functions $I_k(t)$ and $O_k(t)$.

Models using flows alone, being designed for the analysis of general or partial interdependence, have been highly developed and achieved great popularity. They have been used to tackle key areas of economic theory such as, for example, price determination and the net product distribution. Notwithstanding, the way they deal with the time element could be improved. In this respect, Georgescu-Roegen's model combining funds, flows and time,

1. Permits a more detailed analysis of the organisation (character of the elements involved and timing) of the productive processes.
2. Recognizes and sets up an analytical representation of the difference in the way the elements of the production process make their contribution, some of them being agents of the transformation of others.

Georgescu-Roegen saw his proposal as a contribution to resolving what he considered as the serious deficiencies of the production function model. His attention was specifically focussed on the problems of *identifying* production processes: the excessively simplistic treatment of the factors of production.

In a general view, the alternative model of Georgescu-Roegen presents the four following methodological features:

1. It maintains the one-way outlook on the production process (from inputs to outputs) typical of models of partial equilibrium. Consequently, the production conditions for various goods are considered independent of each other, that is, without taking into account the productive interdependences, even though certain conditions relating to the supply of the factors and the demand for output may be added to the model.
2. It insists on a painstaking inventory of all the elements that participate in the production process. Such attention to detail accounts for the time dimension becoming one of the analytical pillars of the model: it is a basic ingredient for classifying the elements of production as funds or flows. The other criterion, as already mentioned above, is the relation-

ship between the time dimension and the boundaries of the production process. There is no doubt that where the funds and flows approach shines brightest is in its ability to describe the timing arrangements of productive activities.

3. It rejects the typical conventionalism of the neo-classical models. Thus, for example, no assumptions with respect to the intrinsic productivity of the factors are added.[18]
4. It ignores significant theoretical developments, rooted in the concept of production function, such as the microeconomic foundations of income distribution and the theory of economic growth. Although with the funds and flows model, Georgescu-Roegen (1972) investigates the new configuration and scope of key concepts such as the marginal productivity of the factors or the casuistic of optimisation, it must be acknowledged that this was not the main thrust of his work. In effect, such issues were hardly mentioned at all in his later work. Indeed, more than on the *ultimate results of the conventional model,* Georgescu-Roegen's disagreement focussed on its basic conceptual design and methodological attributes, both in terms of the priority issues to be analysed and in the way they must be tackled.

Therefore it is clear that the funds and flows model is *not* a mere formal variation of the extensive family of production functions, as would be the case, for example, with the Constant Elasticity Substitution (CES) format with respect to the Cobb-Douglas function, even though it does share certain methodological features, such as the lineal concept of the production process, or that of putting aside the sectoral interdependency.

Despite the improvements his model brought to the microeconomic study of the production process, Georgescu-Roegen was very well aware that it still contained certain serious shortcomings. Indeed, with respect to the funds, it did not satisfactorily deal with the loss of efficiency of fixed capital assets during the course of production activity, along with worker fatigue, and with respect to flows, the role of the harmful by-products that

[18] It might be added that to reject assumed intrinsic productivity means denying the existence of *primary* production inputs. If this attribute is assumed, it would pertain to the funds. In effect, all plots of land had to be, in their day, cleared. Then, on a more or less regular basis the boundaries have to be rebuilt and weeded and the access ways kept clear and usable and so on. If the estate has irrigation, the number of control and maintenance operations increases rapidly: the land has to be levelled, irrigation ditches and drains dug, etc. On the other hand the labour fund requires all sorts of on-going attention: food, shelter, training, etc. Tools and machinery are the output of earlier production processes and require maintenance and repair.

may be produced alongside the main output. To tackle such issues he established three important assumptions, namely:

1. The efficiency of fixed capital assets is held unaltered throughout the successive production cycles.
2. The workers will not analogously experience a loss of their capacity for work.
3. It is suggested to ignore the impact of harmful by-products given that, as is argued, they have no value.

Although Georgescu-Roegen attempted to justify himself, his explanations about these assumptions were too cryptic and evasive. They were so unconvincing that they only added to the confusion. And, moreover, they led to the entire model falling into disrepute. Nevertheless, after a critical review of the arguments given by Georgescu-Roegen, it is not difficult to obtain a clearer understanding of the weaknesses of the model: *all of them come from the limitations of the partial equilibrium approach*. This is the theoretical framework that sustains the funds and flows model.

With respect to the first two points, in the case of classically-based models such as that of Sraffa, the output (multiple) contains possible products and by-products, as well as the units of capital assets worn out and the surface area of land used. However, tired workers do not appear as part of the joint product. The position that Georgescu-Roegen took with respect to the representation of fixed funds and human work was:

By definition a used tool must be an output of some productive process. Yet in no sense can we say that is the aim of economic production to produce used tools. (…) Moreover, with the exception of used automobiles and used dwellings, no used capital equipment has a regular market and, hence, a price in the same sense in which new equipment has. Used equipment, therefore, is not a commodity proper, and yet no report of a productive process can be complete without reference to it.

To be sure, in practice one may adopt one of the numerous conventions used in computing depreciation. But such a solution, besides involving some arbitrariness, is logically circuitous: it presupposes prices and interest rate to be already given. Economic theory has endeavoured to avoid the Gordian knot altogether by building its foundation only upon a process in which *all capital equipment is continuously maintained in its original efficiency*. The idea underlies Marx's diagram of simple reproduction as well as the neo-classical concept of static process. But not all its analytical snags have been completely elucidated. (Georgescu-Roegen 1969: 509, underlined in the original).

And he continues:

To transform our illustrative process into a static one, we should include in it the activities by which the spade is kept sharp and its rivets, handle or blade are replaced when necessary. These activities imply additional labour power, additional inputs and, above all, additional tools. And since the efficiency of these tools must be, in turn, kept constant, we are drawn into a regress which may perhaps stop only after the whole production sector has been included in the process at hand. Moreover, if strictly interpreted, a static process must also maintain its labourers – i.e., the variable capital of Marx or the personal (human) capital of Walras. Thus, in the end, we have to include the consumption sector as well –a glaring illustration of the (...) analytical difficulties of the concept of partial process. To avoid the regress to the whole economic process, we may assume without fear of being unrealistic that in every partial process part of the equipment and all human capital are maintained by outside processes, each one in turn to be analysed separately. After all, analysis cannot proceed without some heroic abstractions at one stage or another". (Georgescu-Roegen 1969: 509-10).

In this extensive quote Georgescu-Roegen combines very distinct ideas[19], which should be examined in parts. To start with, it acknowledges that there is a solution to the problem of how to treat the funds, especially fixed assets, although for reasons hard to accept he rules out used machinery: used equipment is not a *true* commodity as it has no regular market. Even though it may be accepted that there is no production process whose aim is to produce used equipment for sale, there is a second-hand market for all means of transport, from bicycles to boats and aeroplanes, and also a market for most used agricultural and industrial machinery. And, in the worst of all cases, there is always the scrap market. Obviously, the reason for selling used equipment is to recover part of its initial value. Moreover, it is worth noting that even in those cases in which there is no second-hand market, there are still book values. Looking, for example, at a building that has been burnt, the experts will still establish a value for the plot and for the remains of the building.

With regard to the possible alternative of depreciation, Georgescu-Roegen blows hot and cold. Initially he successfully explains why it is not a satisfactory option (the conventions used inevitably include a degree of arbitrariness), although in the end he accepts a kind of subterfuge (the constant level of efficiency postulate) which, as he is perfectly well aware, is not at all defensible. Notwithstanding, as insurance he backs it up with authoritative references. Later though, the overall tone of the argument improves and he refers to the limitations of a partial analysis as opposed to the scope of the general interdependence approach. Unfortunately, the argument ends in the worst conceivable way: appealing to heroics. But at the beginning of the text quoted above, does he not acknowledge the existence

[19] A large number of the arguments given are later repeated in Georgescu-Roegen (1971).

of a definitive solution? Then, what is the sense of such heroics? In fact, it would have been sufficient to delimit the methodological validity and theoretical scope of the model proposed.

Elsewhere Georgescu-Roegen writes:

A simple glance at the activity inside a plant or a household suffices to convince us that efforts are constantly directed not toward keeping durable goods physically self-identical (which is quite impossible) but toward maintaining them in good working condition. And this is all that counts in production (1976: 64).

But, are the expressions *maintaining the funds in their original efficiency*, *maintaining their physical self-identity* and *maintaining them in a good working condition* not all synonymous?

To unravel the matter does not seem an easy task at all. For example, authors inspired by Georgescu-Roegen, use expressions such as *identical state, economically identical form,* and *physically identical form* (Gaffard 1990: 97). However, it is possible to find another meaning: the assumption of the constant original efficiency of the funds can be understood as maintaining the number of fund units and assigning them to the same services (or tasks). This explanation is quite acceptable as it suggests that, with a set of machines operating in the same process, the amortised units are progressively replaced by new machinery. Needless to say, the assumption made rules out the possibility of a machine replacement for economic reasons prior to its physical demise.[20]

On the other hand, in Lager (2000) it is pointed out that the notion of the perennial maintenance of original efficiency should be interpreted as constant efficiency. Thus,

the concept of efficiency involves both the flows of future outputs *and* the flows of future inputs, and therefore, a machine is of constant efficiency if, and only if, the flows of outputs and inputs are constant over its entire lifetime (Lager, 2000: 246, underscored in the original).

This strict interpretation demonstrates the dissatisfaction caused by Georgescu-Roegen's position. Indeed, he treats fixed capital as if it were *Ricardian* land, meaning that its theoretical price is determined by the present value of rental rates paid its use, that is,

$$p_m = \frac{\rho}{r}$$

[20] If efficiency decreases or fluctuates, the question of the optimum cut off time arises: it is important to determine the best moment to withdraw it from production, even though it may still be in a good physical condition. This issue does not arise if efficiency is constant or always on the increase.

p_m being the theoretical price of the machine, ρ the constant perpetual rental and r the rate of interest. Such an interpretation is confirmed by Georgescu-Roegen himself:

> Ricardian land provides the clearest illustration of the concept of fund, but a machine that is continuously maintained and repaired also fits the definition. Accordingly, a machine coming out of a process is a fund even though it may have no part whatsoever in common with the "same" machine that went into the process (Georgescu-Roegen 1976: 41, inverted commas in the original).

The way Georgescu-Roegen deals with the issue of fixed capital explains, as we shall see further on, why the economic dimensions associated with fixed funds (initial price, rates of depreciation and interest rate) are taken as given, i.e., considered as variables exogenous to the model.

Elsewhere Georgescu-Roegen states that to include used machinery and the fatigued worker in the list of outputs of the production process would lead to an erroneous interpretation of production activity. In his own words:

> Our entire analytical edifice would collapse if we were to accept the alternative position that the aim of economic production is to produce not only the usual products but also tired workers and used tools (Georgescu-Roegen 1976: 64).

However, the reality is very different given that including the funds as outputs contributes to improving the understanding of the production process: the funds are not physically incorporated into the output and their productive capacity extends beyond a single production cycle. In reality the objective connection between a fixed fund (a machine) and its services may be established by the following conceptual scheme (Barceló 2003: 21):

$$\text{new machine} \oplus \text{maintenance} \rightarrow \text{used machine} \oplus \text{productive services}$$

This compact production line takes account, on the input side, of the new machine plus the maintenance and spares services required to guarantee maintaining a production capacity similar to the original for as long as possible. On the output side, there is the old machine plus the hours of productive service it has rendered cycle after cycle, year after year. According to this scheme of things, the concept of *preserving the capital intact* could be understood as the ability of the fixed fund to generate practically identical productive services through successive cycles of production. Or, put in another words, the quality and quantity (per unit of time) of the services rendered by the machines, are the same as when it was first installed. Such production services should be qualified as *normal*. As known, the efficiency of fixed funds declines over time. Hence, a progressively intensive degree of maintenance and part replacement operations is needed.

Consequently it is logical that the costs of maintenance and part replacement operations will increase *with respect to previous periods*.

The loss of productive capacity of a fixed fund may take many different forms. One of the most significant is the deterioration of the materials of which it is made due to friction, abrasion, rusting and chemical attack, or erosion. In the case of metal components, lubrication is the standard form of protection. Plastic, on the other hand, will deform or degrade from exposure to harmful agents, etc. Any such damage may affect certain parts of the equipment more than others: take, for example, the tyres of a car, or cutting and polishing tools. All of the above will then give rise to operative problems such as:

1. A higher consumption of input flows (energy) per output unit.
2. Shorter periods of satisfactory operation among successive breakdowns. Increasingly frequent interruptions to service (with the consequent opportunity costs), the solution of which will, moreover, involve the direct cost of the repair work.
3. A slower pace of operation meaning that the production cycle will take longer.
4. A greater volume of by-products or waste products such as swarf, sawdust, and so on.
5. More frequent production of output units that do not meet the quality standards set by the company.
6. A combination of the above together with many other possible functional problems.

Figure 11 shows the hypothetical case of the relationship between the depreciation and maintenance and repair costs of a new vs. an old machine. In the figure, these two machines have the same or similar capacity, as technical change has not been taken into account. Moreover, for the sake of simplicity, the charge for depreciation is constant. Maintenance and repair costs increase with the passage of time, growing progressively as the functional wear of the fixed funds is assumed to accelerate over time. In other words, the loss of efficiency depends on the accumulated number of production cycles in which the considered fund has participated. One of the machines (the older) was installed at t=0 while the other (the new one) starts its operational life at t'. The price of the newer machine is higher in monetary terms than that of the machine installed earlier. Hence the higher charge for depreciation. The top of the chart shows the amortisation and maintenance and repair curves separately while in the graph at the bottom, these are added together. This last graph shows instant t^* as the moment at which the costs associated with the older machine exceed those of the newer unit. Thus the figure shows the process of the replacement of a fixed fund due to the decline of its productive service: any attempt to maintain its level of efficiency intact requires an increasingly higher outlay on repair and maintenance operations which, added to the cost of amortisation, will in the end determine that the fixed fund must be replaced by a new one with lower running costs.

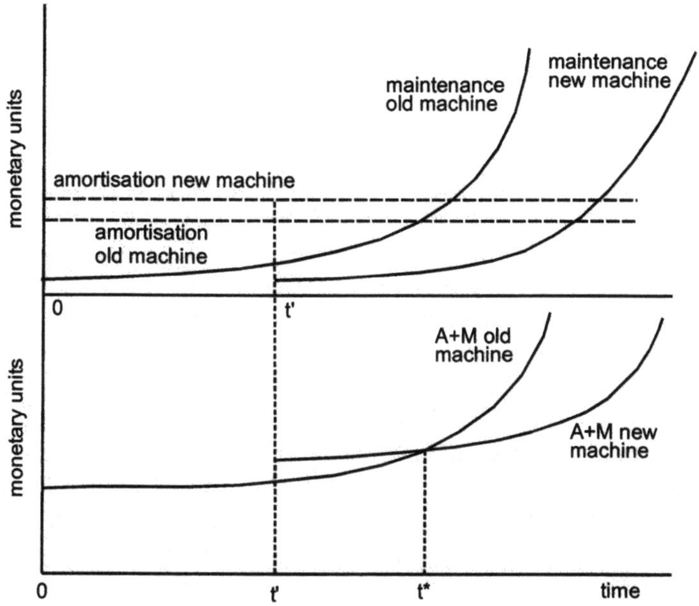

Fig. 11. Amortisation and maintenance of fixed funds

If nothing more, this simple analysis of the processes for renewing fixed funds should be qualified by the following:

1. There are physical barriers and economic restrictions that prevent the total recovery (i.e. as-good-as-new) of the productive efficiency of a fixed fund. The feasibility of keeping a machine perpetually in perfect physical conditions is something that is questionable. In addition, the cost of attempting to do so would prove prohibitive.
2. The technical life of a fixed fund comes to an absolute end in time: a machine is unable to render its services any more and its repair is difficult (for the high cost of rectifying the technical problems observed – such as a light bulb in which the filament has broken-, for the lack of a supply of spares or skilled specialists). The fixed fund then becomes mere scrap. In general, this extreme tends to be very distant in time, given that many machines will get a second chance: although problems such as a loss of efficiency (and obsolescence) have led to these being

withdrawn from the main production process, the services they offer are still of an acceptable quality and so they are kept in the plant as reserve production capacity. They could also be sold on the second hand market.

3. It must be remembered that, in everyday life, the decision to replace fixed capital tends to be made with respect to the production line as a whole. Thus, the essential co-ordination or coherence of the production capacities of the different units (to avoid bottlenecks and excess capacity) may make it advisable to replace a unit when it has only deteriorated slightly.

To summarise, Georgescu-Roegen's proposal takes an interest in the collateral processes required to maintain the productive capacity of fixed funds, assuming that their efficiency may be fully recovered after each productive cycle. Efficiency is kept constant. Such simplification is acceptable given that the funds and flows model is one of partial equilibrium, restricted to very limited scope and time horizons. Otherwise, following the recommendations made by Georgescu-Roegen, maintenance and repair operations should be described in the fullest detail: number of man and machine-hours used and the flows consumed in the form of spares, filters, lubricants, coolants, water, paint, grease, and so on.

The above reflections do not conceal the fact that, in the funds and flows model, the fixed capital is clearly given less consideration than in the neo-Ricardian model, based on the Torrens/von Neumann/Sraffa rule. On the one hand, the latter suggests that new and used machinery should be treated as different classes of goods and, on the other, that the periods of life of fixed capital be considered as different lines of production. This then means that the annual charges of depreciation can then be *endogenously* calculated. The value of this variable is obtained simultaneously with the prices of the other merchandise and rate of profits (wage), given the *numéraire* and salary rates (rate of profits). Moreover, such an approach enables the optimum economic life of fixed funds to be determined, whether these are machines or living beings acting as fixed capital (Barceló and Sánchez 1988; Kurz and Salvadori 1995: 186-218, 2003).

In the case of the labour fund, Georgescu-Roegen's point of view coincides with that of all the other types of economic models of production. In effect, the shared position is not to include the care, training and sustaining of the work force into the model because of its character, both social and economic. For this reason, these activities are considered as exogenous. However, this does not exclude the possibility, if necessary, of such items being reintegrated into the sphere of production.

With respect to waste by-products, it must be admitted that this is an attribute determined by the economic system as a whole. It must not be forgotten that innovation may change the economic role played by a given output. It may be transformed from a useless and hazardous by-product into a basic economic resource, as was the case with petrol.

With respect to the recommendation that Georgescu-Roegen makes, i.e., it is best to ignore by-products because they are valueless the following clarifications should be made:

1. All by-products are, necessarily, produced jointly with another output.
2. The waste from one process may be the input of another process, and consequently such a use will give it value.
3. As a general rule, it is impossible to determine whether a by-product is waste material or not when only the process that generates it is considered.

Any product jointly produced may only be attributed with a price, positive or otherwise, from the point of view of the economy as a whole. In other words, the issue of the price of by-products may not be resolved without bearing in mind the demands of the reproductive requirements of the system. Only if by-products were not to be used by any other process of production or consumption, including recycling, would they then come to be considered waste. The cost of waste disposal could be interpreted as a negative price.

In conclusion, the funds and flows model cannot deal with by-products properly, a feature shared by any other model belonged to the partial interdependence or equilibrium approach.[21]

---

[21] It must be said that an attempt has been made to incorporate the funds and flows approach into a model of general interdependence (Tani 1988). Given of the lack of further research in this area, it is difficult to determine whether or not such an analytical work holds promise.

2 Productive deployments of elementary processes

The most outstanding feature of the funds and flows model is its attempt to offer a detailed description of the production process, especially its time dimension. For this, the theoretical development of the fund-flow model on the following pages will give priority to this dimension. Namely,

1. Working with the characterisation of the different ways of deploying an elementary process, as proposed by Georgescu-Roegen, the relationship between the technical division of labour and the optimisation of production times will be analysed in depth. This is an area that has already been tackled by the economists of the Classical School, since it underlies the increasing yields of manufacturing activity through the specialisation of workers. However, benefitting from this gain will require an acceptable level of synchronisation of the different consecutive production tasks, a problem which may be satisfactorily tackled with certain conceptual offshoots of the funds and flows model in close relationship with some operational research tools.
2. The impact of the time dimension on plant costs will be carefully assessed. In particular, short-term cost curves will be developed in which the time factor plays a relevant role.

The analysis will also include typical high points of microeconomic production theory, such as the indivisibility of fixed funds, economies of scale, process innovation, or whatever. However, in contrast to the conventional model, the suppositions guaranteeing *convenient economic behaviour,* such as convexity or diminishing returns will be abandoned. This implies losing concepts previously considered essential, such as the substitution of factors, marginal productivity and so on.

The funds-flows model pays particular attention to the forms of deploying the productive activity, understanding this as the way of executing a stream of elementary processes. There are only three basic forms of deploying them: sequentially, in parallel or in line.[1] Their essential features are:

1. Sequential deployment consists of performing an elementary process after the other. In this way, once one output unit has been completed the production operations corresponding to the next unit are started. This method of organising production is expected to be found in craftwork activity. For example, when making a piece of jewellery, a goldsmith will do all the tasks from start to finish, before manufacturing the following.
2. Parallel deployment is typical of agriculture. In this case, production involves a large number of elementary processes, all carried out simultaneously. In effect, if each plant or tree is considered as a separate process, a farm may be seen as an extent of land in which the same type of elementary process is repeated many times.
3. Line production, commonly known as mass or series production, is the industrial form of producing goods. The advantages of this form of production were finally acknowledged at the beginning of the 19th century, when it was known as the *Factory System* (Babbage 1971; Leijonhufvud 1989; Landesmann and Scazzieri 1996b). In this case, the units are produced in such a way that, there is no need to wait for the previous unit to have been completely finished to start on the next one. For example, long before a car leaves the production line the assembly of the following unit has been started. In reality, it is not just *one*, but *many*, subsequent units that are already on the conveyor belt. It should be pointed out that although all industrial activity can be adapted to line production, there are several clearly differentiated sub-types (batch production, continuous flow production, etc.), depending on the pace of the process and the amount and variety of output units produced per unit of time (Spencer and Cox 1995).

[1] This triple division was first proposed by Georgescu-Roegen (1969, 1971, 1972, 1976). Since then it has been upheld by many other authors: Tani (1986), Landesmann (1986), Wodopia (1986), Morroni (1992), Scazzieri (1993) and Gaffard (1994). It should be pointed out that the term *sequential* has been chosen here for the modality that Georgescu-Roegen, and his closest followers, called *series* activation. The reason for this conceptual change is to avoid confusion, given that in both common and academic terms, *series* manufacturing is the expression used to refer to industrial transformation in general. Conversely we have maintained the use of the words *line production* to refer to industrial production, as proposed by Georgescu-Roegen. Thus the term *series* production will not be used in this text.

This chapter will offer a detailed examination of all the different ways of organising the production of goods. The goal is to identify the distinctive features of each different form clearly and to delve deeper to identify the pros and cons of each system in terms of efficiency.

2.1 Sequential production

To start with, the sequential production system is characterised by the consecutive activation of elementary processes, that is, one after the other and always unit by unit. Only when all of the phases of the previous process have been fully executed will production of the following output unit start. There is no doubt that sequential production is the simplest way of deploying an elementary process. Figure 12 shows the sequential activation of an elementary process with duration of [0, *T*].[2]

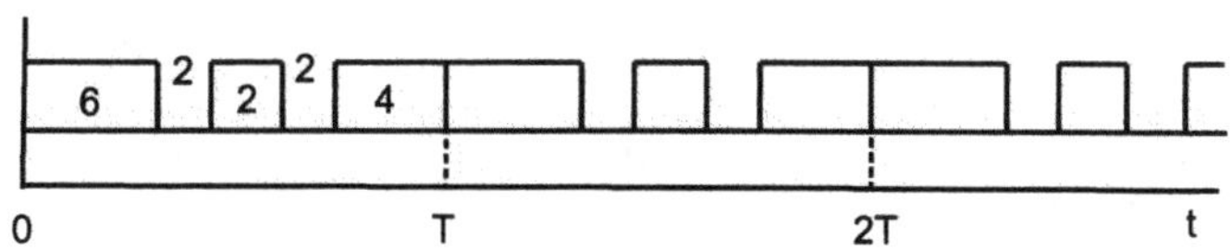

Fig. 12. Sequential production

The above sequential process, with a duration of $T = 16$ hours, comprises three intervals of activity (of 6, 2 and 4 hours) and two periods of inactivity (each lasting 2 hours). These periods of complete inactivity could have been dispensed with, but their inclusion makes the example more generally applicable. With the above data the level of production per unit of time (1 hour) is 1/16 output units. In general, the level of production is $1/T$ output units.

Although it may be applied to any production process, sequential production characterises artisan activity as well as the construction of major transport infrastructure such as bridges, tunnels, and canals (Georgescu-Roegen 1986: 257). Moreover, it may be used for the assembly of certain means of transport (ships, aircraft) the main attribute of which is their size and/or exceptional and unrepeatable technical features.

---

2 For the sake of simplicity it has been assumed that the intervals of activity of the funds will coincide between them and, moreover, inflows are not represented. It has also been taken that the output flow will emerge at the end of the elementary process (instant *T*). Thus, in reality, the figure shows the duration of the different phases (or tasks) of the elementary process.

The problem with sequential production is that the fund, or funds if several work together, will unavoidably be subject to periods of idleness. Indeed, the services of a fund are no longer required after its task has been done. Therefore, it will remain inactive until the next production unit when it will again take up its work. More often than not, the situation may even arise that virtually all of the funds remain inactive, as can be inferred from figure 12 above. For instance, this will be the case for the craftsman who is waiting for sun-dried bricks.

There is *no* solution to such idle periods when elementary processes are deployed consecutively. An alternative to try to mitigate the inefficiency of the sequential deployment is to extend the working life of fixed assets, and the working day of the operators. However, such a solution does not avoid the problem of the down times of the funds between the different tasks/phases of the elementary process. Any real solution must involve new forms of deployment of the elementary processes.

Prior to concluding this section it must be remembered that in a sequential process, in which the human work fund is very important, as tends to be the case with craft production, the total duration of the elementary process (or direct labour time per output unit) is highly variable, given the flexibility of the labour fund with respect to changing the pace of operation.

2.1.1 Changing the sequential process

Although the structure of sequential production processes is very simple, any modification to it warrants a certain degree of analysis. Obviously assessing the differences between two processes producing exactly the same output, but by slightly different methods, might involve the comparison of an extensive list of elements, tasks and times. In any event, it would be interesting to compare the total process time, the mean time of application of the flows, and the degree of use of the different funds.

With regard to the first point, the reduction (α) of the duration of the elementary process, that is,

$$T=(1-\alpha)T, \text{ with } 0<\alpha<1,$$

may be expected either by a reduction in the intervals of activation of the fund or by lessening its periods of inactivity or, obviously, a combination of the two. Thus, the level of production per unit of time will rise from $1/T$ to

$$\frac{1}{(1-\alpha)T}$$

For example, taking the data shown in figure 12, it is clear that with the elimination of the idle periods, the elementary process is reduced to $T=12$ and therefore $\alpha=25\%$ which gives rise to an expansion of the production per hour to 1/12 output units [$1/(1-0{,}25)\cdot 16$]. As has been recently indicated, behind of this fact may probably be changes in the yield of labour, which is a decisive factor with respect to the duration of the elementary processes deployed in a sequential manner.

With regard to the second point, without denying the interest there may be in comparing the total amounts of each flow consumed, it is also important to look at their average application time. With a single number, a good approximation of the time profile of the amounts of a given inflow could be achieved. This indicator is defined as follows (adapted from Frisch 1965),

$$\frac{\sum_{\tau=0}^{\tau=T} \tau \cdot f_\tau}{\sum_{\tau=0}^{\tau=T} f_\tau}$$

where,

1. τ ($\tau=0, 1, 2, ..., T$) are the equal periods of time into which the elementary process could be broken up. For the sake of simplicity, the time path of the flow has been divided into discrete intervals.
2. The term f_τ represents the amount of the inflow in period τ assuming, for the sake of simplicity, that the flow is incorporated at its beginning.

The indicator of the mean period for the incorporation of a flow in an elementary process normally changes because of variations in the amount of inflow along with changes in its time profile. In order to distinguish between modifications in the amount from those in the time profile, a detailed knowledge of the $F_i(t)$ function of the flow is essential.

With respect to the degree of use of the funds, this may be simply defined as the ratio of the service time to the total duration of the elementary process. Taking the data from the example above, assuming the work fund provides its services in all the phases of activity, the degree of use would be 75% (12/16). Clearly this ratio would change according to the time of services and idle times.

To sum up, whatever indicator is chosen to examine changes in the temporal profile of a sequential process, knowledge of the specific time path of the funds and flows is required in order to have a complete understanding of the causes. In general, this information is laborious to obtain and perhaps bothersome to interpret. It should not be then surprising that the microeconomic theory of production prefers to ignore the time factor. Nonetheless, the extra cost of using a model with an explicit time dimension is more than offset by the extra analytical depth achieved.

2.2 Parallel production

Parallel deployment comprises the simultaneous activation of different consecutive elementary processes so that the phases follow the same calendar. In other words, the elementary processes start and finish together, all (or most) fully overlapping each other. See figure 13, which represents a parallel production process for four elementary processes, each with exactly the same pattern of activity and idle times.

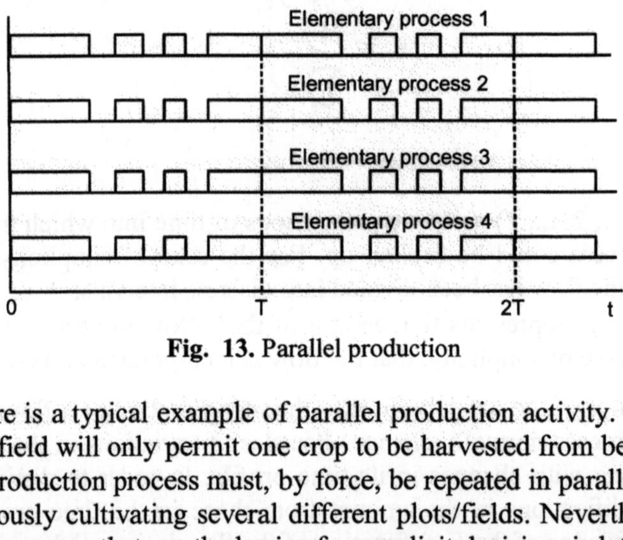

Fig. 13. Parallel production

Agriculture is a typical example of parallel production activity. Given that each plot/field will only permit one crop to be harvested from beginning to end, the production process must, by force, be repeated in parallel. That is, simultaneously cultivating several different plots/fields. Nevertheless, any sequential process that, on the basis of an explicit decision, is laterally replicated in the same production unit also constitutes a parallel process. For example, one single facility with a number of dry docks may have the capacity to simultaneously construct different units of the same type of ship.

With respect to sequential activation, a parallel arrangement may be seen as an attempt to achieve the objective of making more efficient use of the funds. However, it is evident that this is not the case. Down times will still persist. Only when a fund is capable of operating on several different elementary processes at the same time will parallel deployment improve the general efficiency of the process. In effect, instead of intervening in one single elementary process (or current output unit), the fund will perform the same production operation for several.[3]

Refining the analysis, there are two main ways of organising parallel production:

1. Processes in which the time scale is set for physical, chemical or biological reasons.
2. Processes which tolerate the modification of their internal operating times.

The difference between the two depends on the degree to which the different phases (or tasks) of the parallel process overlap in time. In the case of the former, there is a total simultaneousness while, for the latter, there is the possibility of delaying, at will and individually, one or more of the phases (or tasks) of the elementary process. So, in this case the condition of matching is rather more relaxed.

2.2.1 Parallel process with rigid time schedule

In mid-latitudes, farm production in the natural environment is still a strictly parallel process. Indeed, the use of the reproductive possibilities of animals and plants is limited to their specific biological rhythms. Phases (or tasks) of those processes have a notably rigid order and distribution over time. Operations such as tilling, sowing, watering, calving, or shearing occur currently at strictly given times in the overall production process. In reality, even though some degree of tolerance may exist, there is not much room for change. At most a few hours or days, which it is scarcely significant with respect to the total duration of the elementary process.

In the case of annual crops, there may be single phases (or tasks). These are activities that once executed will not be repeated until the appropriate point in the following process next year. In farming, there are also shorter cycle phases (or tasks) such as milking. All these cycles are ruled by circumstances, like the climate or the day-night cycle, beyond human control.

Output takes the form of a periodical flow. In the case of vegetable products, the pace of the output is discontinuous and reaches its peak at the

3 A furnace in which hundreds of bricks are fired is an example of a fund capable of simultaneously executing many elementary processes. Thus, firing is deployed in parallel. Later this issue will be dealt with in greater depth.

time of harvest (often an annual event). At the same time, the amount of output is uncertain and will depend on the whims of the weather. In the case of livestock, batches of product may be harvested more frequently (Georgescu-Roegen 1969; Zuppiroli 1990; Romagnoli 1996a; Polidori, 1996).

The above features are not so marked in tropical farming, where the constant nature of the climate throughout the course of the year permits sowing and harvesting at practically any time. The process does not necessarily have to be replicated constrained by seasonal rigidities and therefore a continuous output flow may be maintained (Georgescu-Roegen 1969: 531 and 533). However, by keeping environmental conditions constant at a suitable level by artificial means (greenhouses, stabled livestock), many species (both animal and vegetable) may sustain high rates of growth and/or reproduction without the environment influencing their biological cycles. Such production is called industrial or better milieu-controlled intensive farming.

Whatever the climate of the region or the technology employed, agricultural production will always require the funds to provide their services intermittently, and for varying lengths of time through the different stages of the production process. Much machinery will remain inactive for long periods as the task in which it is involved is activated once, or at least, not very frequently, during the course of the production process, depending on the stage of plant growth and/or appropriate atmospheric conditions. The tooling used to prepare, sow, combat weed, harvest, and so on will only be used a few days or weeks a year. Even though funds are used as fully as possible in these short periods, they will always remain inactive most of the production process. On the one hand, all the above hinders the specialisation of funds. Indeed, the more specific a machine is, the less time it will be used for. This could lead to thinking of farming as over-mechanised process: a large array of specific tools combined with little use of them. On the other hand, this under-use gives rise to a bigger depreciation burden. Coupled with all these difficulties, the problems related to the scattered nature, the irregular geometry, and small size of the plots must be added. In any event, the aforementioned inconveniences should not represent a severe economic setback for farms, even if they are of small extent: agricultural equipment comes in all sizes and at different prices, and the more expensive items tend to be bought collectively. In the final instance, such equipment may always be rented.

In terms of time, there is no doubt that the greatest continuity of use is made of the human labour fund, and the fund which offers the service of traction (animal or machine). In this later case, as crops are planted over large areas, there is a need to transport all kinds of auxiliary machinery.

With respect to the Ricardian land, it is a fund which is always present. If the farm is small this insufficient size combined with the cyclical nature of the harvests justifies the farmer lamenting the loss of potential earnings if she/he does not have any alternative form of occupation. However, this does not mean that she/he is considered unemployed: simply on hold. To all events, the owners of smaller farms do tend to opt to engage in diverse activities, given that the issue of the periodical inactivity of the funds cannot be satisfactorily resolved by replicating parallel processes. It must not be forgotten that the idea of extending the area of land cultivated clashes both with the high price of land or very unattractive rents to pay. The question is to adopt a strategy that intensifies the traditional diversification of crops. To seek new sources of income, be it on the farm itself with, for example, integrated livestock breeding not linked to the cycles of the herbaceous crops of the farm, or outside of the farm by way, for instance, of paid temporary employment on other farms or in other productive sectors. Such multifaceted activity may involve one or several people on the farm.

Nonetheless, despite the somewhat particular nature of farming, diverse innovations have been introduced:

1. The duration of one or more of the phases of the production process has been reduced by accelerating certain tasks.
2. A decrease in the volume of services required from the funds, man or machine hours.
3. A reduction in the amount of flows required per output unit.

Technical changes to be considered are of two kinds: changes to the number of elements of production, or to the time for which the productive funds operate. Obviously, both may also be combined. In any case, these innovations have permitted the farmer to harvest more land, increase the yield per hectare and divert the man or machine hours thus made available to other activities. The urgency to do so will depend on the economic scale of the farm, a variable which it is mainly related to its extent and the quality of the land.

Without a shadow of doubt, when attempting to analyse such changes in the organisation of agricultural processes, the conceptual framework proposed by Frisch (1965) is of the greatest interest.[4] Having acknowledged that taking time into consideration enhances the concept of the amount of a factor, the author explains that the application of the labour input (or fund)

---

[4] One of the referees considers these concepts from Frisch as an anticipation of Georgescu-Roegen's model, although the former did not pay particular attention to the time dimension of productive activities. Indeed, he proposed to build production functions considering the process as instantaneous.

to the same production operation may, in two or more farms, present different time profiles, even when the same total amount of human work is employed. Figure 14 shows therefore two different input curves relating, for example, to the number of man-hours per week (vertical axis) required to pick fruit over a period of several weeks (horizontal axis). The two curves, each of a different thickness, refer to different farms.

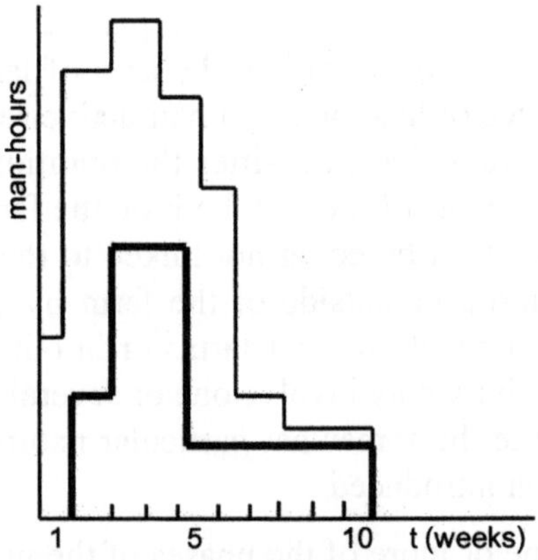

Fig. 14. Input curves

The area enclosed by each input curve indicates the total amount of labour used. It is clear that on comparing the time profiles for the labour factor applied to fruit picking, one may be led to suspect that there are differences between the methods of cropping used by the two farms. In effect, assuming the same total output, a lower number of man-hours per week may reflect the fact that the harvesting process is more highly mechanised. As a general rule, changes to the input curves may be accounted for by changes made to production procedures. For example, another case would be if the application of a given factor were to be accelerated. This could be represented by a curve that would be narrower at the base and higher.

Another contribution made by Frisch was his phase diagram, an extension of the input curve concept. An example is given as figure 15. In this figure, with *x* and *y*-axes representing time and the amount of a given element required respectively, it is shown a phase diagram, which provides three different kinds of information:

1. Each input curve encloses an area equivalent to the amount of the input applied.
2. It indicates the number and order of applications of the fund or flow element throughout the production process.

3. It shows the pace of successive applications, given that they are clearly dated.

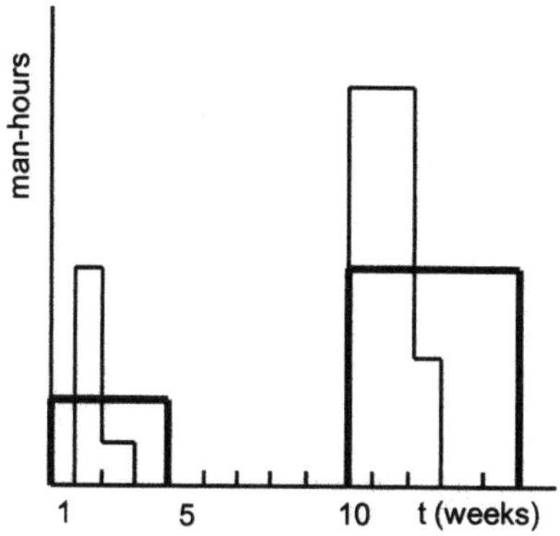

Fig. 15. The phase diagram

A phase diagram allows the foundations to be established on which to build a broader typology of process innovation for the primary sector. Without attempting in any way to be exhaustive, it could be distinguished:

1. Innovation that permits the performance of a given task (or phase) to be accelerated, without any variation being made to the time at when it occurs in the process. Thus the fields will be ploughed at the corresponding time, although with advanced equipment, the task will be completed far quicker. The hours of service of one or more of the funds will fall per output unit produced. To all events, the total duration of the production process will not diminish.
2. Technical improvements that lead to a reduction in the number of interventions required by the crop: an herbicide that with one application eliminates all weeds for the whole growing season.
3. Innovation that permits a reduction in the amount of a given flow required per output unit, i.e., the use of a drip irrigation system.

The concept of phase diagram enables us to identify and precisely assess the impact of any technical innovation made to agricultural production processes.

2.2.2 Non-rigid parallel activation

Despite the above with regard to agriculture, a parallel process could be imagined in which the phases are susceptible to postponement, within broad margins, and at will. In other words, the historical time taken between two consecutive phases may be extended as desired, without endangering the viability of the process. Although the phases of the elementary process generally occur in a strictly sequential order, there is great flexibility in terms of the time that may be taken between them. The reason for this is that there is no physical limitation to the total duration of the elementary process. Depending on the technology used, *real* duration does not necessarily have to be the same as the theoretical *minimum* duration. Such a form of activation could be called parallel production with *tolerance* of the phase (task) timing of the elementary process. An example of this is given in figure 16.

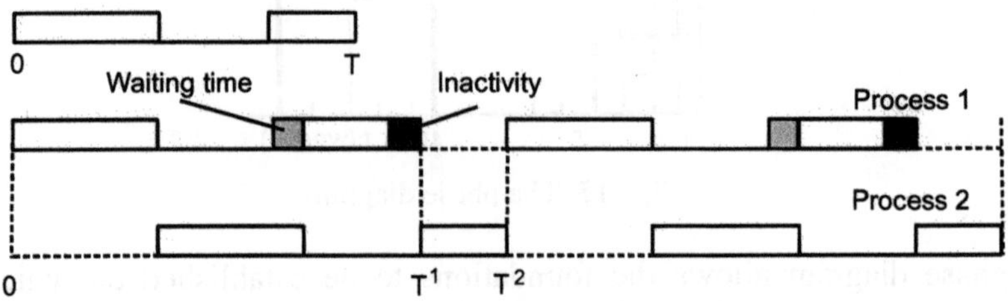

Fig. 16. Non-rigid parallel activation

The above very simple case takes an elementary process which employs one single fund (U^*) which is operative in two phases, separated by an idle period which cannot be reduced, but may be extended. In effect, in order to reduce the idle time of the fund, another elementary process is activated out of step with the first, leading to the second phase of the original process being displaced. Thus, a waiting period is added at the beginning as well as an interval of idle time at the end. This way, the fund U^* *jumps* from one process to the other, meaning that the idle time will be less than if it were operating on one single elementary process at a time.

Given the greater continuity of operating time pursued by jointly activating a set of different elementary processes, it is logical to expect that the degree of use of the funds, measured as the ratio of their intervals of activity and the total duration of the elementary process, will be increased. Consequently there are endless possibilities with respect to the internal composition of the times of activity and inactivity in such a process (Tani 1996).

The example shown is termed parallel production because of it always uses the same fund unit and simply multiplies the number of elementary processes until achieving the greatest synchronisation of the phases possible to ensure the fund remains idle as little time as possible. However, the aim is not for the phases of the elementary process to be perfectly synchronised, which is, as we will see below, a basic requirement for line deployment. The situation depicted may occur, for example when building housing or executing public works. Often a builder will perform several similar processes slightly out of step, or reorganising the phases, with the explicit aim of reducing the idle times of the funds (manpower and machinery) as much as possible. For instance, a building firm that is urbanising a large urban area from the initial levelling of the ground through to the laying of basic utilities (water, power, sewers and so on), organises its work so that the different phases are performed consecutively across the whole area, without forgetting that the work may well have to be synchronised with other projects in which the firm is engaged in other nearby places. The result is that, for a specific site, the time the work takes is extended beyond that which would be strictly necessary. Nonetheless, from the point of view of the company, the funds will remain idle for less time as, after executing a task in one place, the activity there is halted while the fund is taken to another area to execute the same task.

2.2.3 Parallel vs. functional process

One should not confuse parallel deployment with the well-known stratagem commonly employed by small workshops: to avoid breaks in the use of certain funds, several *different* elementary process, but sharing the same sphere of activity, are kept active at the same time. This is known as the *functional* or *job-shop process*, the main features of which are the following:

1. The different elementary processes have common singular tasks. These tasks have a flexible duration.
2. Both the human work and capital equipment funds are capable of performing a more or less broad spectrum of tasks. The flexibility of arrangement of the phases demands that a single operator or instrument should be capable of productive interventions of varied content. The funds must be characterised by a certain degree of versatility.[5]
3. The link between the different tasks (or phases) in one single elementary process is functional. There may be discontinuity between them: the breaks in work do not imply the funds being truly inactive as they will be employed on other elementary processes. They jump between stages of different processes.

---

[5] Certain funds are characterised by outstanding versatility. Such is the case of many universal tools: a hammer, a screwdriver, scissors, etc. Of course such versatility is limited. A knife may serve to peel potatoes or to make a fruit salad, while a saucepan is used to boil or stew different kinds of food. However, neither of the two can substitute the other.

This is a form of organising production activity which prevails in those sectors in which the demand is discontinuous and requires short runs. Such would be the case, for example, of a restaurant, tailor's, hairdressers', or home appliance repair shop.

Figure 17 shows the participation of two mechanics in a small car repair shop. They work in different elementary processes arranged as a functional scheme.

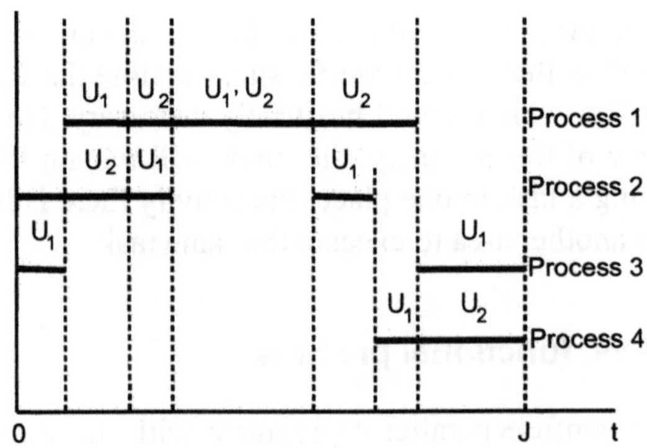

Fig. 17. An example of functional organisation

In this hypothetical workshop, over the course of the working day (J), 4 elementary processes, corresponding to the same number of different activities, are carried out. The first, for instance, consists of repairing a vehicle, with the corresponding phases of dismantling and replacement of the faulty component, and reassembly. Elementary process number 2 is a large-scale repair job, with multiple phases which are performed with great discontinuity. The third elementary process includes the replacement of a part that is not in the workshop store and will not arrive until a few moments before closing. The last elementary process consists, for example, of a client who is in a great hurry and wants an immediate oil change. As can be seen in the figure, both mechanic 1 (U_1) and mechanic 2 (U_2) move from one phase, or part of this, to another in either the same, or a distinct elementary process, with no loss of continuity. At times, the two are engaged in the same process. This is unavoidable given that, as we are looking at different processes, the timing of the different stages need not be coordinated. Moreover, the execution of the elementary processes may be interrupted for many reasons.

In short, a functional process is a particular variation of the *sequential* process. That concept would actually be a more appropriate term for what occurs in an artisan workshop: elementary processes with clearly different outputs, although all pertaining to the same family of goods, are performed with a greater or lesser degree of time overlapping.

2.3 Line production process

Line production processes are characteristic of, though not exclusive to, industrial plants. They consist of the progressive activation of elementary process of the *same* type, but with a certain delay between them. Thus, given that the lag is of a shorter duration than the elementary process itself, the successive elementary processes overlap each other. Figure 18 shows a process with duration of [0, *T*] and two tasks (or phases) separated by an interval of inactivity.

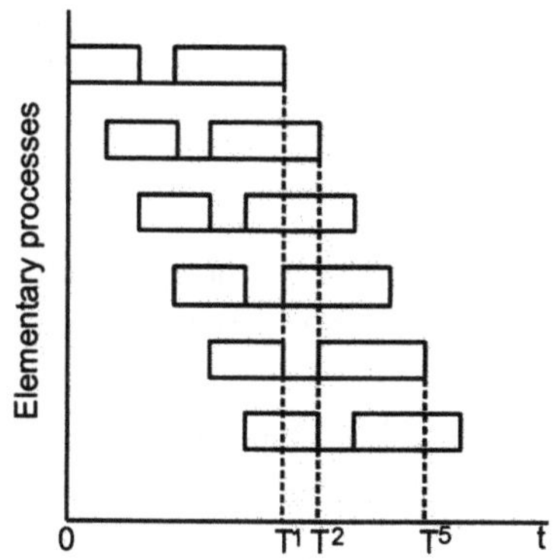

Fig. 18. Line production

In line production, a new elementary process begins regularly after a short lapse of time. An output unit is then obtained after every brief period of time: at points T^1, T^2, ..., T^5 of the figure. The distance in time of between these points is less than the duration of the elementary process. Thus, in an industrial plant one will observe, at any given moment in time, multiple output units in production, each in one of the different consecutive phases of production.

One general condition required for an elementary process to be performed in line is that the moment of starting, and finishing, should not be subject to any seasonal rigidity. The manufacturing process is, of course, the most pertinent example of this. Industrial manufacturing processes produce many different kinds of previously designed standardised artefacts and substances, on a large scale. The resulting product tends to be made of interchangeable parts of precise and invariable dimensions. There is much room for manoeuvre when seeking to rearrange production operations to find the least costly, provided that the outcome is still technically viable.

Manufactured products tolerate the line assembly of their components, i.e., once a task (or phase) in the assembly of an individual product has been completed, the same operation may be performed on the next unit in the process, without having to wait for the whole production process of the previous unit to finish. Once a certain task has been performed on an output unit, the next unit appears without delay, ready for the same intervention. This permits both machine and worker to specialise in a single uninterrupted task (welding, painting, or whatever). In this way the funds located at a given workstation repeat the same production operations time and again.[6]

Line production tends to be associated with access to extensive markets. In effect, compared to sequential or parallel processes, line manufacture multiplies the amount of merchandise produced per unit of time. That is why the term *mass production* is used when referring to line production. It must be pointed out that this process entails employing a great number of funds, and hence the far greater physical scale of industrial, as opposed to artisan, facilities.[7]

It is worth noting that, while one cannot precisely measure the degree of fund capacity used in the functional process, this is not the case with line process. The specificity and regularity of the tasks performed facilitate the establishment of procedural patterns along with the control of the operating times of the funds involved. In the case of the labour fund, this way of thinking is known as Taylorism.

The company consultant F. W. Taylor (1856-1915) developed and disseminated methods of work control, the aim of which was to maximise the yield of line processes. To these ends, he proposed fragmenting the different phases of production and reducing them to tasks that would be extremely simple to execute. Then, he recommended optimising the micro-movements of the operator at his/her workstation. In order to achieve such ends, the smallest movement was timed, any unnecessary displacement eliminated, and standard execution times established for production opera-

---

6 Although not explicitly stated, it is assumed that the funds may move from one consecutive process to the next at either no cost, or at a cost lower than that of inactivity.

7 Hence the common perception of the scale of production as something physical (the size of the facility) while it would be a more pertinent to consider the scale as the volume of output per unit of time. This issue will be dealt with in greater depth later.

tions.[8] Taylor also proposed improvements to the design of tooling, the quality of the materials used, etc. The basic idea was to constrain the worker so that his/her production would be invariable. It should be pointed out, however, that Taylorism showed relatively little concern with human resource selection and management. Issues such a motivation and conflict resolution remained hidden in the background.

Taylorism placed traditional work rules in doubt. Indeed, the control of time and motion established the strict separation between the design of a process and its execution, two aspects which were inseparably bound in artisan production practice. Through applying Taylorism methods, management drastically reduced the degree of discretion afforded to the workers when performing their work. This was previously guaranteed by the traditional workshop codes of conduct.

It should be said that Taylorist practices are specially fitted to the line activation of the production process, but not necessarily associated with any particular form of technology. Line processes, as we will see more specifically in the next chapter, provide enormous possibilities for increasing the productivity of work. However, along with such latent promises, poor functioning of the line brings on significant economic damage. A right and proper cooperation of the workers is crucial to avoiding line shortcomings and to prevent huge economic losses.

2.3.1 Line vs. parallel deployment

Depending on the characteristics of the elementary process to be performed, two forms of line activation may be distinguished:

1. Unitary, as in the generic example given in the figure above.
2. The simultaneous deployment of multiple elementary processes.

The deployment of several elementary processes executed as a block leads to a delicate terminological problem. The problem is rooted in the degree of similarity that there is between the elementary processes in question.

---

8 However, Taylor was not a pioneer in such timing. It must be remembered that the first chronometer was built in 1690. Seventy years later, criticisms appeared about the installing of clocks in workshops. At the beginning of the 19th century, James Watt Jr. pioneered a time and motion study. Babbage, on the other hand, recommended controlling times without the knowledge of the workers, and took the mean of the observations made at different times of the working day as his significant variable. What was really new with Taylor, and shared by his disciples, was his insistence on the *scientific* nature of such practices.

Despite the ambiguity of the boundaries, the following three levels may be defined:

1. Elementary processes that bear only a slight resemblance to each other. Although a given task (or phase) may be shared, and/or there are funds that may be used in all of them, the elementary processes (and the goods produced) are different, even though they may all fit into the same main category. This is the case discussed above of *functional* organisation, which is applicable to establishments producing goods or offering services of a clearly job-shop nature.
2. Elementary processes of great similarity. In such a case, the term *related* outputs is proposed. This is the situation in which the elementary processes correspond to goods with only slight differences. For that reason, all of them belong to the same, single and restricted range of products. As a rule, only certain flows and tasks will vary, meaning that the output units from different processes may share most of the productive operations. The existence of such shared interim elements and stages is translated, on the one hand, into a better uptake of the capacity of the funds and, on the other, to expanding the degree of diversification or scope of output. Typical examples of this are the refining of petroleum to obtain different fuels, or the production of several dyes in a chemical plant. Some of these outputs require either the incorporation of specific flows or the services of particular funds.
3. *Identical* elementary processes, replicated in parallel. As will be explained in chapter 3.4, such situations are associated with the use of *indivisible* funds, the productive capacity of which must be exhausted. In that case, the level of production could be multiplied without an equivalent multiplication of costs. The line activation of a single kind of elementary process repeated in parallel breaks with the concept that the funds sited at the different workstations have to work on only one *single* output unit in the process. To the contrary, many processes use funds that operate with multiple capacities. An example of this could be machines that simultaneously paint, cut, fill or assemble several different units of output. Within the framework of the model developed herein, the indivisibility of funds is thus reinterpreted as the quality of being able to operate simultaneously on more than one elementary process. This cannot be obviously said of inflows.

Care must be taken not to confuse *related* production and batch production of *different* outputs (or elementary processes). As is known, batch production is concerned with diverse goods belonging to the same range. Such would be the case, for example, of a garment manufacturer. Of far greater importance is the fact that the uniform batches are executed one after another. There is no *simultaneous* production (or in parallel) of the different lots on the same production line. The use of batches is normally justified by the nature of the demand: to cover different segments of the market and to adapt the current production swiftly to constant market changes.

Related production, and by extension, that of parallel deployed, may benefit from economies of scope, understood in the strict sense of the term, i.e., referring to the production process as such and not to the plant or firm. The fact that different products are sharing one or more common inflows or funds accounts for their simultaneous production is being more economical than if they were produced separately (Bailey and Friedlaender 1982). Formally,

$$C(q_1, q_2) < C(q_1, 0) + C(0, q_2)$$

C being the cost and q_1, q_2 the outputs from elementary processes performed in parallel. As indicated above, one possible source of common production cost saving is the fuller use of the capacity of indivisible funds.

As one may observe, the concept of economies of scope incorporates a degree of ambiguity: it is defined without considering time and then inadvertently mixes parallel (in its diverse forms) and batch production. In other words, the concept does not clearly distinguish between an output which results from the *simultaneity* of very similar elementary processes (performed in parallel) and a *successive* diversity of products that emanates from a periodical change of batch in the process. This is not a denial of the relevance of the concept of economies of scope, but does make it hard to specify the reasons behind such economies.

Before closing this section, it is important to note that what has been called related production should not be confused with obtaining several different outputs from one *single* elementary process, that is when one or more main outputs are accompanied by different waste or by-products, as well as possible emissions in the form of smoke or dust particles. This coexistence of different types of outflows is what is generally expected from real production processes (Steedman 1984).[9] There are two main types:

1. Strong or binding association: The ratio of the amounts of the outflows obtained is highly stable. Such is, for example, the case of grain and straw, and of many other by-products with respect to the main product.

[9] We have taken great care not to call the coincidence of different outputs flow *joint production* so as to avoid confusion. In effect, the term *joint production* also tends to be used to include the capital used between the process outputs. Given that the model developed herein only places the fixed funds amongst the inputs, it has been considered preferable just to speak of *types* of simultaneous outflows.

2. Weak or separable association: Altering the rates of certain inflows or the service times of given funds, the proportional relationship between the different outflows may be significantly altered. A classical example of this is the different ways in which sheep are fed, depending on whether they are bred for meat or wool. Another example is the way some animals are fed in order to cause the hypertrophy of certain organs or tissues, as is the case with the production of *foie-gras*.

If, for the sake of simplicity, it is assumed that there are just two outflows, a graphic representation as in the figure 19 is obtained.

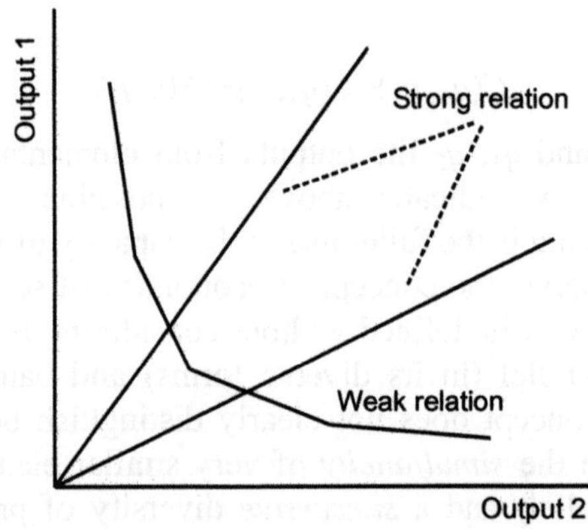

Fig. 19. Types of simultaneous outflows

In this hypothetical example, the radial axes indicate a strong association between the outputs, while the convex segments represent variety of separable outflows. In the latter case, the specific pathway of the *trade off* of the different outputs may vary greatly.

2.3.2 The Factory System

The precise origins of the *Factory System* are not known. In fact the Factory System is a modification of the artisan form of producing goods which permitted the workshops to meet a progressively growing demand per unit of time. There appear to have been rudimentary forms of the Factory System developed in Western Europe in the first half of the 13th century, when there was an abundance of projects to build great cathedrals and cloisters. Contemporary representations of the activity of the masons tend to suggest that they would have specialised: some in the cutting of the blocks of stone and others in sculpting. Another better documented predecessor is to be found in the great arsenals of the 18th century. They produced arms using a line deployment (Mumford 1963; Heskett 1980; Best 1990). However, it was at the beginning of the following century that the Factory System came into its own and was established as an innovation and powerful form of producing manufactured goods (Scazzieri 1993; Grahl 1994). Certain authors at that time believed the line organisation of production was one of the most important advances ever made by humanity (Babbage 1971). Under deeper scrutiny, it was more of an organisational than a technical innovation (Georgescu-Roegen 1971: 248, 1976: 68-9).

The dissemination of the Factory System, and the improvements it brought with it, was accompanied by reflection on the technical division of labour and times of work. In effect, while the *Encyclopédie ou Dictionnaire raisonné des Sciences, des Arts et des Mètiers* (1751-65) subdivided the manufacture of watches into 21 operations performed by different workers, Babbage stated in 1834 that, with the increasing fragmentation of work, the number of operations involved already exceeded one hundred. There is no doubt that the manufacture of needles was the most widely expounded process as an illustration of the progressive levels of specialisation. A pioneering example is Perronet's *L'art de l'épingler* (1762). This work contains numerous detailed explanations of the manufacturing methods, times and costs for such goods. Following the trends of this time, Adam Smith also illustrated the advantages of the technical division of labour with the same example. Moreover, he added that its degree of deployment bore a direct relationship to the extent of the market.

Despite the importance that Smith himself gave to the division of work, which appears as the frontispiece of his main work, the issue was dealt with in greater depth by the British mathematician and engineer Charles Babbage.[10] This author was fascinated by the productive capacity of the great industrial plants: his analysis focussed on the relationship between the technical division of labour and the scale of production operations. The main theme of his argument was that the pursuit of a fuller use of the funds, that is, the reduction of the periods of inactivity of both man and machine, had led to the activation of processes in line, with the corresponding specialisation of the funds and the emergence of the great facto-

[10] References to the life and work of Charles Babbage (1791-1871) can be found in Berg (1980 and 1987) and Stigler (1991). There is also Babbage's autobiography (1969). It must be added that his contributions to political economy are nearly forgotten today (Rosenberg 1994). In general, economists have paid very little attention to Babbage's economic writings. By way of example, Schumpeter (1954) merely devotes a footnote to him.

ries. As reiterated in Babbage (1971), the great industrial plant was clearly superior to smaller scale production: the price of the merchandise manufactured was comparatively lower.[11] Indeed, this big unit of production:

1. Permits the maximum deployment of the technical division of work and, with the simplification of productive operations, facilitates the introduction of time saving machinery.[12] In this way, productive operations achieve enormous scale (or high output level per unit of time).
2. Concentrates all the different stages of manufacture under one single roof, thus reducing the cost of transport of semi-manufactured goods and facilitating the control of worker activity.
3. Permits the selection of those workers whose skill profile best meets the needs of the task in hand, thus leading to increased worker performance, especially in those cases in which payment is by piecework.

With respect to the first of the above points, Babbage recognised that the division of labour was an inexhaustible source of increased yield. His preference for grouping workers together in a large factory suggests that the author was highly aware of the limitations of the *putting-out system*. Disperse cottage industry, although not preventing the technical division of labour, made it hard to realise its full potential in terms of productivity. The reason was the problem of coordinating the increasing fragmentation of the process. In order to achieve that, it would be better to gather workers in a single space.[13] In addition to this logistical problem, the state of the roadways at the time and the inclemencies of the weather also have to be borne in mind. In short,

The transition form *job-shop* and early *putting-out system* to modern forms of manufacturing organisation was based upon the identification of three fundamental sources by which productive efficiency may be increased, that is, (i) increasing the degree of task differentiation and task specialisation, (ii) emphasising the continuity and linkages between the different stages of the fabrication process, (iii) re-

[11] Babbage was not the first to observe the increasing yield of great industry. Antonio Serra, an Italian economist born in Cosenza at the close of the 16th century had already commented on the phenomenon.

[12] According to Babbage, the durability of the machines depended more on the regularity than the speed of operation. With respect to maintenance, he insisted on the need for proper lubrication (Babbage 1971: 8).

[13] Without forgetting, as Babbage highlights, there are two types of machine. One has the task of transmitting power and executes its work while the other is the source of energy. The latter represent such an investment, and generate so much power, that they are only viable for the larger industrial facility.

solving problems of utilisation of indivisible fund-input elements and processes (Landesmann and Scazzieri 1996b: 270, italics in the original).

After the triumph of the large industrial facility, the main interest of the entrepreneur turned to the control of working times, that is, demanding punctuality and repressing absenteeism.

The third point cited in the previous page was ignored by Adam Smith. On the contrary, Babbage wrote:

That the master manufacturer, by dividing the work to be executed into different processes, each requiring different degrees of skill or of force, can purchase exactly that precise quantity of both which is necessary for each process; whereas, if the whole work were executed by one workman, that person must possess sufficient skill to perform the most difficult, and sufficient strength to execute the most laborious, of the operations into which the art is divided (Babbage, 1971: 175-176, all text in italics).

This is known as *Babbage's Principle*[14]: specialisation allows separate tasks according to the degree of skill or strength required. It is essential to hire individuals whose physical profile and experience best match the task required. The principle is a response to the labour market context at the time: the effective organisation of work was in the hands of the craftsmen, all of whose families would have been hired. The existence of family ties guaranteed a greater degree of discipline, but also meant that the work had to be divided between workers of both sexes and of very different ages. Quite plausibly, this labour market situation could account for the attention Babbage paid to the appropriate skill-based assignation of workers: a highly skilled worker performing a simple task was a wasted resource, while the opposite represented excessive risk in terms of accidents, breakdowns, and so on.

Having said this, it is quite evident that the British engineer's concept of training was somewhat crude. In the first place, qualification included both regulated and tacit knowledge (acquired through experience), and being appropriately predisposed to the work. This is far more complex than Babbage's concept. Secondly, the author was unaware that all workers, *at the same time,* have to adapt to the specific working conditions of the post they occupy. If not, they must be replaced. In short, Babbage's postulates should be reinterpreted in the sense that, at most, specialisation permits the refining of the criteria of staff selection (Corsi 1991: 16). To succeed, however, it has many other connotations.

---

[14] The same principle had been formulated earlier by Melchiorre Gioja, an Italian economist, writing in 1815. His contribution is summarised in Scazzieri (1981, 1993: 43).

In the same order of things, Babbage (1971: 170) also pioneered an approach to the economic dimensions of apprenticeship: the training of workers represents a cost to the company and so it is essential to recoup any such investment as quickly as possible. This argument is graphically represented in figure 20.

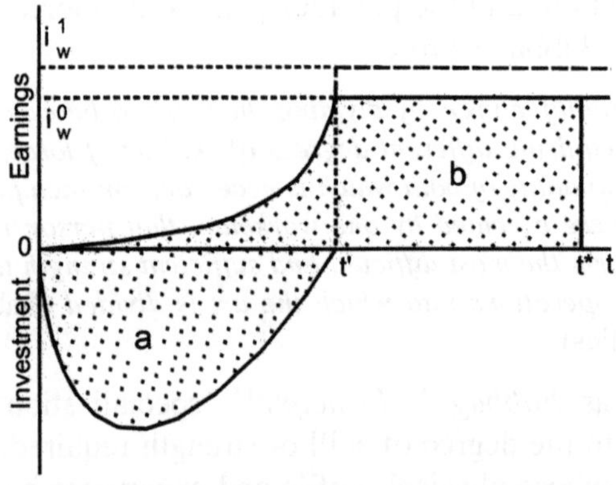

Fig. 20. Training and the production process

The cost of training, possibly higher in the initial stages, is represented by the area *a*. The contribution of the new worker, quantified by attributing part of the total production per unit of time to him/her minus the wages he/she is paid, is represented by area *b*. At instant *t'* the work performed by the apprentice is of an equivalent quality to that of the other workers of the same category, while the company has to wait until *t** to recover the investment it has made in his/her training.[15] According to Babbage, the goal of the company is to reduce the period of recovery of the cost of training as much as possible. One way of achieving this is by simplifying tasks. A greater technical division of labour reduces both the direct cost and time of training.[16] In terms of the figure, area *a* decreases in both directions. Another way is to pay the trainee at under the normal rate. This option may even be extended longer than is strictly necessary. In this case, according

[15] Rosenberg (1994: 30) regards Babbage as a forerunner in terms of the theory of Human Capital. Probably Rosenberg is being somewhat impetuous. While models such as those of Gary Becker, Jacob Mincer and others attempt to explain salary structure in terms of the human capital accumulated by the individual, Babbage merely tackles one very specific issue: how the company recovers the investment it has made in the training, in an environment with established salary rates.

[16] Babbage did not consider the psycho-social aspects of training.

to figure 20, the net contribution of the worker will be above that of the other operators: $i_{w^1} > i_{w^0}$. In this situation, value t^* is closer in time.

On the other hand, Babbage judged that:

When the number of processes into which it is most advantageous to divide [a process], and the number of individuals to be employed in it, are ascertained, then all factories which do not employ a direct *multiple* of this latter will produce the article at a greater cost (Babbage 1971: 212, italics in the original).

The scale of line manufacturing with specialised workers presents a particular form of discontinuity: having adjusted the amount and pace of activity of the workers, which translates into a given output volume manufactured per unit of time, to extend the scale of production, one would need to employ a *multiple* of the existing number of workers (and machines). If this rule is not followed, the unit costs of production will then increase (Landesmann 1986: 308-9; Morroni 1992: 63-5). This is known of as the *Multiple Principle*, and is subject to the following conditions:

1. All phases of the process have the same duration.
2. All possible idle time has been eliminated from the process at the original scale (*well-arranged factory*).

However, it must be pointed out that the multiple's principle is nominal as, even exploiting the possibilities of the technical division of labour to the full, optimisation of the process is never perfect. Deviations may occur as a result of unforeseen circumstances and it must also be borne in mind that companies will normally wish to maintain a certain reserve of production capacity.

Marx (1976) insisted on the difference between heterogeneous and organic manufacturing. An example of the former is the manufacturing of timepieces by numerous independent workshops, each of which makes just one of the components (cogwheels, springs, cases, etc.). In this context, even though the workers do specialise, there is no strict coordination in terms of time between the different operations of the elementary process although, obviously, all the different parts must be available prior to final assembly. In heterogeneous manufacturing, the cost of coordination may be high. On the other hand, in the case of organic manufacturing, all the activity takes place in the same plant. In such a case, the different production tasks are organised like a tree, and are performed in order and with strictly planned timing. The operations are deployed in series (or line, as per the terminology used here) and, in general, it is possible to increase the number of processes activated until all fund idle time is eliminated. This is achieved without logistic problems causing any loss of continuity between the different stages.

Marx established the following general rule with respect to the organisation of production in large factories: if T is the length of time an elementary process lasts and t the shortest time any of the individual tasks takes, ceaseless activity of the funds would require elementary process to be activated in line with a lag equal to T/t. Having optimised the production line, any enlargement of this should be made in accordance with Babbage's principle of the multiple (Landesmann 1986: 298).

For authors such as Babbage or Marx, after establishing the optimal scale, the process of production may only be replicated in *parallel*, that is, by a proportion that may only be expressed in integer numbers. Consequently, after having determined the output level per unit of time that, with a given production technique, will guarantee the elimination of all fund idle time, any attempt to increase the volume of production will imply a fall in the overall efficiency of the process, with significant repercussions in terms of costs, unless the new scale proposed is a multiple of the former. All that is illustrated in figure 21 in which the x-axis indicates *total* production per unit of time. As can be seen, in that hypothetical plant, the same process has been replicated three times (in a parallel layout).

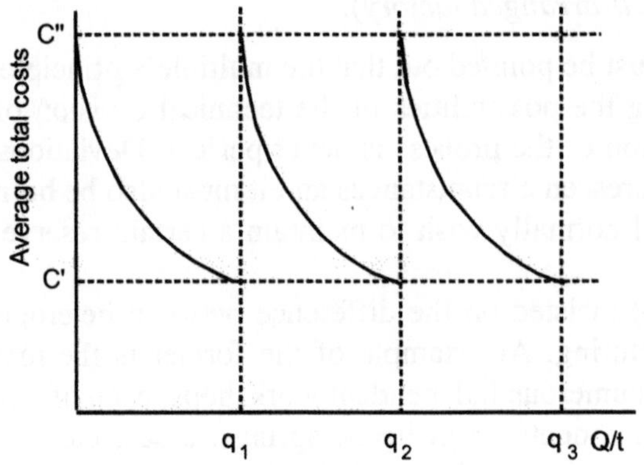

Fig. 21. Costs and the Multiple's Principle

Taking the optimum level of production as q_1, any increase in this will raise unit costs above the minimum level of C', until reaching q_2. This situation is then repeated between this level and q_3. In short, only for multiple volumes of q_1 is the same cost *low* of C' achieved. Thus *efficient* scale will increase *discontinuously*.

Let is consider a very simple elementary process: the manual placing of letters in an envelope. The first step is to adjust operation times. That is, none of those who separate and fold the paper, those who seal the envelopes, those who stick the stamps on and, finally, those who stick the address label to the envelope are ever kept waiting. So, it is easy to see that any attempt to increase the number of finished letters is economically inefficient, although technically feasible, unless the whole of the optimised line is replicated properly. Indeed, if one line produces 30 letters per minute, producing 40 would require another line to be set up, which will work well below the efficiency optimum. Until it reaches an output of 30 letters/minute, the new line will always suffer losses in terms of time. In few words, what it is required is several optimised lines working in parallel: the *total* output achieved will be a multiple of the initial output of the first optimised process. The same is true of the number of individuals involved and, if applicable, the machines used. In short, greater scale does not necessarily mean greater efficiency (or lower unit costs). Evidently, technical innovation will cause optimum levels of production to vary. However, the multiple's principle will still be applicable.

At the end of the 60s, Georgescu-Roegen returned to many of the earlier ideas of Smith, Babbage and Marx and applied them in a more exacting formal context. Thus he definitively established the advantages of the line deployment of a given elementary process. Namely,

1. The disappearance of the intervals of inactivity of the funds.
2. The possibility of systematically increasing the output per unit of time.
3. The unstoppable advance towards greater levels of specialisation and mechanisation.

Nevertheless, it must not be forgotten that,

> in many respects, the classical authors made division of labour central to their analysis of manufacturing forms or production organisation. However, they were unable to actually determine appropriate degrees of fund-input specialisation, apart from emphasising that this degree should be highly sensitive to the overall scale of production (…) (Landesmann and Scazzieri 1996b: 279).

In reality, what was left pending analysis was far more, as Georgescu-Roegen and his followers have stressed. In effect, there was a lack of discussion about the general conditions for the deployment of line processes and its outstanding theoretical implications. These issues will be tackled in the following pages. However, prior to that, a look should be taken at the different forms of line production.

2.3.3 Forms of line manufacturing

In the world of industrial production, line deployment has become the standard for the general organisation of production processes. To all purposes, there are several variants on this, depending on the degree of standardisation of the output, the flexibility of both the process and funds employed, and the layout of these funds. The resulting diversity constitutes the so-called *Strategic Manufacturing Theory*. This is an attempt to group together the enormous plurality of methods and technologies of industrial production in just a few categories.

The Strategic Manufacturing Theory classifies the different types of line process according to two main criteria:[17]

1. The first dimension refers to the degree of intermittency of the production *stream*. The possibilities range from a flow which may be adjusted at will to the production of a set output amount which is continuous over time.
2. The second criterion is the degree of similarity of the successive output units manufactured by the plant. This will run from each unit having its own particular physical and/or functional features to a situation in which all the units produced are absolutely identical, i.e., without the slightest difference among them because of a maximum level of standardisation.

Figure 22 provides a comprehensive overview of the basic forms of industrial production. The different types are listed diagonally: from the top left-hand corner, where the level and features of the output are easy to adjust to the requirements of demand, through to the bottom right-hand corner, where the volume and homogeneity of the flow are hard to adapt to a changing market. Thus, the output flow progressively gains compactness, in two senses: through the progressive similarity of the output units produced, considering normal plant activity, and the greater difficulty, for technical and/or economic reasons, of completely interrupting or altering the amount produced per unit of time.

2.3.4 Jumbled production flow

The jumbled flow or job-shop, consists of the manufacturing of unique articles or, at most, of limited runs of a few types of products which all belong to a single family of goods. This requires the use of highly flexible funds, given that the kind of output produced constantly adjusts to the changes in demand. Obviously, in this case, the workforce will be made up of specialists, whose experience in the trade will be worthy of the highest consideration, while the tools used will be universal. Here, of course, the

[17] Developed from Wild (1972: 3-18), Cusumano (1992) and Spencer and Cox, (1995).

form of production will be *sequential* and, strictly speaking, this is not a line flow process. Therefore, it is shown on the chart merely for the purpose of comparison.

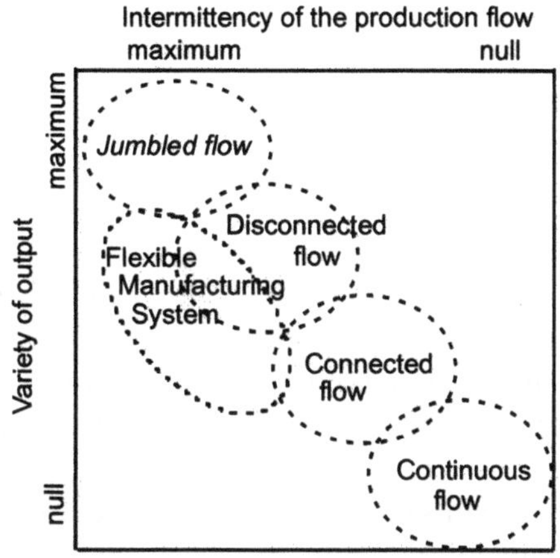

Fig. 22. Forms of line flow (or industrial production)

2.3.5 Batch production and the flexible manufacturing system

A disconnected flow corresponds to what is known as batch production, or the line production of alternating groups of products all from the same range, that is, relatively different but standardised products. Such may be the case, for example, of the production of dolls, the printing of books, the manufacturing of furniture, the production of components and simple parts, and so on. In this kind of process, production is programmed in line according to the demand cycles. Such programming will include the scheduling (or determining of the times and methods), dispatching (or definition and assignation of the tasks to the funds), follow-up (or coordinated movement of materials) and the control of the results.[18]

Apart from that, in such processes the fixed funds must be regularly reprogrammed to adapt to the peculiarities of the new output to be produced. Logically, the larger the batch, the lower the repercussions such reprogramming will have on costs. Likewise, the shorter the *switching time,* or idle time between batches, the better it is for the firm. In batch production, the existence of a general table of batch switching times for i ($i = 1, 2, ..., I$) products as shown below, is suggested (Piacentini 1997: 174):

[18] For batch production management, see Kitsner, Schumacher and Steven (1992), McKay and Buzacott (1999) and Hennet (1999).

	1	2	...	i	...	I
1	0	t_{12}		t_{1i}		t_{1I}
2	t_{21}	0		t_{2i}		
...						
i	t_{i1}			0		t_{iI}
...						
I	t_{I1}			t_{Ii}		0

Each term on the table indicates the switching time the process funds require for a change of output. As is clear, the table need not be symmetrical. Moreover, this table of switching times could be accompanied by another/others relating to the flows and fund services required for the switching of the batch.

Technological or organisational innovation could reduce the magnitude of one or more elements of the table, and it could be interpreted as increased process flexibility. Although a quantitative indicator of the additional flexibility achieved could be designed, it must be acknowledged that flexibility has many facets and is a concept with powerful qualitative connotations.[19]

A typical example of a batch process is the manufacturing of machine tools. Here, standardised components are produced in batches whose machining stages (turning, milling, etc.) can be automated. With the appearance of numeric control units, with programmable memories, it has been possible to reduce batch size while maintaining an equivalent level of costs. The modules are later assembled to configure the different machine tools constructed, a task which frequently requires additional adjustment, both to guarantee the proper functioning of the common components, and to ensure the correct fitting of the unique elements. Therefore, the process of work usually requires the services of highly skilled human work.

On the other hand, flexible production systems comprise sets of machine tools and robots deployed in production cells, connected by automatic conveyor systems for the output in process and for any other materials that may be required. The entire system will be computer controlled. The whole purpose of such systems is to process, at low cost, relatively small batches of parts or components. Highly versatile equipment is required for this, that is, it will be reprogrammable so as to be able to cope with the execution of multiple different operations. Thus, the variety of

---

19 The section 6.6 deals with the issue of flexibility.

parts that may be manufactured and the total number of process routings is increased.

Flexible production has its roots in the USA at the beginning of the fifties, when the first numeric control machine tools were built for the military aerospace industry. In 1969, the first truly flexible production system appeared, combining machine tools with computer guided vehicles and computer assisted design. Since the sixties, the advances in robotics and AI technology have led to the development of an enormous variety of such systems (Freeman and Louçã 2002: 312-3).

It must be pointed out that the enormous complexity of flexible systems greatly increases the cost of their successful implementation. In effect, the largest part of the implementation costs is taken up by the software and supplementary instrumentation and facilities. Moreover, each generation of machines (or robots) has its own specific capabilities and special tooling, meaning that reprogramming is both complex and expensive. Likewise, in terms of reliability, flexible production suffers from the insufficient development of automatic error detection technology. Consequently, such flexible production systems have not been able to achieve equivalent unit production costs for batches of highly varying sizes. Working with a limited variety and relatively long production runs guarantees a better uptake of the economies latent in flexible production systems (Archibugi 1988; Olhager, 1993).

Finally, it should be added that flexible systems represent a skills loss for the labour force as the work may often be reduced to the feeding of certain machines (loading and unloading operations), supervision and daily routine system maintenance operations.

2.3.6 The moving assembly line

Connected flow production covers everything from the manual assembly line to the transfer line. Whatever the case, discrete output units, comprising exchangeable parts, are produced on a conveyor belt or other automatic system of conveyance.[20] The output in process moves through the different

[20] Given that line activation means continuous activity and specialisation, there are different specific forms of developing a line process. One example would be where the operators moved from one static output unit to another. Another is the conveyor belt, which moves boxes containing the production output and different components to be assembled. Here, the individual worker will merely add a specific element and return the output to the box ready for the next intervention (Coriat 1982). Other forms of conveyance include truck lines, overhead hook lines, etc.

workstations, identified by the content of their work, i.e., by the execution of a set of specific production operations using certain tools and facilities. Such lines are also known as *Repetitive Manufacturing* lines:

Repetitive manufacturing is the production of discrete units in a high volume concentration of available capacity using fixed routines. Products may be standard or be assembled from standard modules. Production management is usually based on the production rate (Spencer and Cox 1995: 1282)

The production of home appliances, cars or electronic consumer goods are all examples of one kind or another of connected flow production methods.

The most genuine of such production systems is the Moving Assembly Line. It first appeared in the mid 19th century at the Colt arms factory in Springfield (Massachusetts), and at the Chicago slaughterhouses. The definitive demonstration of its full potential, however, did not come until Henry Ford implemented it at his Highland Park (Detroit, Michigan) Automobile plant in 1913 (Nolan 1994).

The manufacture of interchangeable parts or components was another innovation that facilitated assembly line production, although for many years, such interchangeability encountered enormous difficulties. This was basically owing to the lack of standardised design techniques and measurement instruments, in addition to sufficiently precise machine tools. Despite the improvements made with respect to the number and degree of standardisation of components throughout the nineteenth century, the so-called *American Manufacturing System* remained a hybrid system, with some parts that were interchangeable and others produced manually. In doing this, it was impossible for many years to completely assemble any device with randomly selected parts and guarantee that it would operate properly (Heskett 1980).

Finally in 1908, Ford managed to achieve a high degree of interchangeability for his automobile parts. He used specialised and high precision machine tools, and imposed a standardisation of parts and tools, both in materials and precision. After that, it was possible to eliminate virtually all the rectifying operations that had previously been required during assembly. However, the workers still moved from one product unit to another around the plant (Freeman and Soete 1999: 142). It was not until five years later that Ford at last combined the concept of simplifying assembly tasks as fully as possible with a system for conveying the output in process. His system linked workstations with single-skilled workers. The specialisation of workers in the execution of highly simple tasks facilitates the work supervision as well. This synchronous and continuous conveyance line runs at a speed that management could adjust at will. As a result, the speed of the assembly line increased. Needless to say, these improvements in productivity were achieved after a long period of trial and error. To sum up,

> The Ford system, starting out with the standardization of a product, went on to the standardization of parts and jigs, the development of specialized machines, the fractionalization and consecutivization of operations, the automation of conveyance by means of the conveyor belt and the setting up of an assembly line operation, and the achievement of synchronization of production, until it finally brought to completion a modern mass production system (Shimokawa 1998: 99).

It should be stressed that, when Ford's assembly line system was started up, another target was posed: the drastic reduction in unnecessary work inventories. Indeed, the priority of taking advantage of production scale, through the line's running at high speed in a perfectly balanced manner, gave rise to increasing line rigidity. Then the suitable procurement and delivering of parts became a key factor for permitting a smooth line operation. Stocking big lots of them near each working stations was the solution found.

Ford's moving assembly line permitted drastic price reductions (the Ford T cost $850 in 1908, $600 in 1913 and $360 in 1916) while the other options available at the time were far more highly priced (for example, an electric car cost $2,800 in 1913). Ford took advantage of the economies of scale and the superior technical versatility of the internal combustion engine.[21] As a result, the number of American manufacturers fell from some sixty in 1900 (with a joint annual production of 4,000 vehicles) to half a dozen in 1929 (with an annual output of 4.8 million units) (Freeman and Louçã 2002: 273-280).

Despite the above, it must not be forgotten that the Ford T was a cheap, basic product which was none too reliable mechanically speaking. It was designed with a rural environment in mind, for drivers with knowledge of mechanics and tools at hand. The problems of reliability, more commonly encountered in mass produced goods, represented a cost to the client. After the Second World War, the changes the Japanese made to Ford's production method were aimed, amongst other things, at reducing the number of faulty parts this method produced (Freeman and Soete 1999: 140-4).

Thus, from a business point of view, the *standardised and synchronised system* presented immediate, indisputable advantages, the most important of which are the following:

---

[21] In 1917, the major American manufacturer of steam driven cars, the *Stanley Motor Carriage Company,* produced only 730 units per annum, less than Ford in half a working day. Moreover, the ability to move on all types of roadways and the low cost of the fuel finally tipped the balance in favour of petrol engine cars.

1. A sweeping reduction in the time required to manufacture output units.
2. A drastic reduction in the time required for training. The high level of specialisation facilitates the learning process. This was of special importance when relying on heterogeneous, immigrant labour with no specific skills training.

However, new problems also arose. Firstly, the fragmentation of tasks multiplied the number of workstations which, in turn, had shorter activity cycles, and thus the transfer times gained greater relative importance, while the synchronisation of the whole process became far more complex. Secondly, the working conditions on the assembly line were rejected by the workers. The speed of the line, the boringness of the simple and repetitive work, the ever-present noise and the severe discipline overwhelmed workers. The work had simply become a race not to fall behind. The pace set by the machines and the monotony represented significant physical and mental wear respectively. The result was a feeling of suffocation and oppression. All this account for the high rates of staff turnover: 400% at the Ford plant in 1913 (Freeman and Louçã 2002: 277). Given that any productive system requires the collaboration of the worker, albeit to make the informal adjustments required to temporarily resolve any unforeseen faults, a lack of motivation increases the vulnerability of the line. A weakness aggravated by the consecutive nature of the line: any problem, albeit local, will easily bring the whole line to a standstill. So it was not surprising that, to combat the rejection described, at the beginning of 1914, Ford introduced wages of $5 a day, doubling the rates normally paid at the time.

Ever since, companies with assembly lines have preferred to hire people with very little chance of employment elsewhere (normally newly arrived foreign immigrant workers), made a large part of the wages paid dependant on achieving production targets (around two thirds variable with no great importance attached to length of service), implemented plans to facilitate the rotation of workers around different tasks, created work teams to foster emulation and reciprocal control, and have always attempted to automate production operations as fully as possible.

The third kind of problem associated with assembly lines was their inflexibility with respect to changes in the size and make-up of demand. This is why many important manufacturing sectors have always preferred to maintain more flexible production systems, integrated into local or regional networks of associated firms. Nor did the concept of standardisation immediately convince all and sundry. In effect, right up to the eve of the Second World War, many manufacturers of tooling and machinery were suspicious of it, feeling it would open up the field to newcomers and increase competition in the sector. However, although many consumers rejected the monotonous design imposed by the American manufacturing system, it could also be used to mass produce the smaller component parts of clearly differentiated goods (Sabel and Zeitlin 1997).

To complete the list of the limitations of assembly line production, mention should be made of the sensitivity of the production flow to disturbances caused by unforeseen factors (power cuts, breakdowns, cancellation of orders, or labour conflict), all of which tend to represent significant costs when activity is started anew.

To sum up, the assembly line has as a whole been shown to be rather fragile, given that it requires complex synchronisation, accurate forecasting and the cooperation of the workers to operate at its best. Hence, there is exhaustive control of workers' tasks. Over the years, Taylorist and Fordist methods, combined with techniques to improve production routing (the PERT method of calculating the "critical route"), the universal standardisation and normalisation of components and products and the automation of processes (Coriat 1982, 1993), have all contributed to improving the reliability and efficacy of assembly line based processes. To all events, the most decisive new development was the reform of the Fordist line, known as the lean production system, developed gradually by the Japanese car manufacturers, starting in the fifties.

With a market of limited scope, the protection of trade barriers and the prohibition of foreign investment in the sector, the Japanese automobile firms developed their own particular version of the assembly line, notable for the following relevant traits:

1. The establishment of work teams: the line is fragmented into sections rather than points (one work-station, one worker, one task). Each section has its own associated group of workers whom, in accordance with production targets set by management, distribute the tasks to be performed amongst themselves, thus gaining in flexibility (each worker must be able to perform any of the tasks assigned to the group as a whole) and, most especially, in internal control (the remuneration and various incentives depend on the results of the group as a whole, thus helping contain absenteeism and foster emulation).
2. Associated with the above, we find the so-called *Quality Circles*. These take in the suggestions made by the workers to improve either the process or the product, and seek their ever greater involvement in quality control.[22] In the Fordist tradition, this function was exclusively per-

[22] Non-participation normally leads to the imposing of sanctions. The suggestions made normally regard quality control and improved performance. Management decides which suggestions should be taken up and, if applicable, implemented.

formed by an expert group at the end of the assembly line, without worker involvement (Freeman and Soete 1999: 152).

3. The inventories held along the assembly line are reduced to a minimum. At the same time, stressing measures are exerted on the suppliers. They are encouraged to supply smaller amounts with greater frequency (*just-in-time* -JIT- delivery systems), and thus have to take upon themselves the cost of storing a stock of parts and components. JIT is a system

   (...) able to adapt flexibility to quantitative and qualitative changes in production through communicating to all corners of the workplace the latest production information on production trends as they respond to market trends and shifting conditions on the production line by using notice boards (*kanban*) on which were posted information on when, where, and who has produced how many lots of parts and where the lots have been delivered. Hence, (...) JIT makes possible a flexible management of the production process that attempts as much as possible to adapt to market trends and changes in production methods (Shimokawa 1998: 94).

   Nonetheless, it must be pointed out that Ford was also worried about the size of the parts and components inventories in their plants. But, Ford gave the highest priority to the continuity and smooth running of the assembly line or, in other words, in taking advantage of the economies of scale. In a context of facing a huge demand, the goal of keeping the inventories to a minimum, was put aside.

Although all the changes above explained have been able to break the extreme functional rigidity of the early Fordist system (based on the premise that there is inherent opposition from the worker to any kind of rules of work), both the short production cycles and routine nature of work remain the same. Moreover, the Taylorist division of the design and execution of a task, and the need for time and motion studies to establish procedures, still exist.

A particular case of connected flow line production is that of automatic lines: these consist of different workstations set up with automatic machinery which performs a sequence of operations on discrete parts, connected by mechanical conveyance systems which automatically position the output in process. The earliest examples of such systems date back to the 1920s, although it was not until after the Second World War that their use was to become more widespread in sectors such as the production of cars or home appliances. There are two main types of automatic lines:

Thus, participation consists of the worker making proposals, but others finally decide.

1. Traditional transfer (or synchronised transfer) lines, used for the machining of engine blocks, the pressing of automobile bodywork, and so on.
2. The more recent development of cells made up of numeric control machines and robots, capable of performing relatively sophisticated production operations, such as welding.

Although the tasks performed by automatic machinery are simpler than those performed by a human worker, recent advances in microelectronics and robotics have enabled more versatile systems to be developed (and, in turn, more flexible production systems). However, in the case of the automobile sector, some stages of cars assembly, their final checking and the supply of materials and parts to the workstations, still resist complete automation. Manual work is preferred because of its levels of perception, capacity for anticipation and the celerity in resolving small incidents. In all of these capacities, human faculties greatly exceed those of industrial robots (Morroni 1991, 1992; Coriat 1993).

2.3.7 Continuous-flow process

The continuous-flow process is the fourth of the line production systems to be examined. When the material to be processed is of a fluid nature (for example liquids or semi-fluids, bulk raw material, or whatever) it is feasible for it to be processed without interruption for long periods of time. Such processes use tailor-made purpose built automatic equipment. In such cases, the main work of the employee is to permanently monitor the system. His/her tasks are not defined by where he/she is positioned in the elementary process, but by his/her partial contribution, within a broad range of functions, to achieving the goal of guaranteeing that the production facility operates following the design specifications, that is, according to the economic and safety parameters established.

In the case of the chemical, electricity generation and other industries, the flow of production is managed from a control centre that receives information from a complex, branched system of measuring devices (pneumatic instruments, electronic sensors, etc.) and issues commands to regulating and operational devices.

In a continuous process, devices that automatically interrupt the process when certain quantities deviate from preset values exist alongside verification operations which have to be performed by workers (regular checking of facility control and inspection systems). So, the process is never fully self-regulating as not all possible disturbances are immediately, automatically detected. While with simple machinery many breakdowns are predictable and easily located, in complex automatic continuous flow systems it is impossible to have prior knowledge of all the possible events (faults) that could trigger industrial accidents. What is more, the dense fabric of control points is also, in itself, subject to problems of reliability. The stress of productive systems brings about corrosion or vibration, which also have an effect on the efficacy of the response. It should also be remembered that advances in control systems tend to be applied in environments in which they will be stretched to their limits which, in itself, does nothing to help reduce their own vulnerability.

One important feature of the continuous process is that it is very expensive to stop or disconnect, albeit for a breakdown and/or reasons of safety. Without delving into the quality with which it is performed, maintenance work tends to be constant, and carried out according to established procedures. It involves a small number of highly skilled workers with a great diversity of attributes, people whose lives could be at serious risk in the case of accident, but whose greatest annoyances under normal circumstances are working night and weekend shifts.

3 Characteristics of line process

As shown in the preceding chapter, line processes abound in manufacturing, especially when the output comprises assembled parts or components, such as cars, domestic appliances, and so on. All such processes are defined by the following three features:

1. They consist of functionally independent phases, deployed in a given order in time.
2. The elementary processes, activated consecutively at given intervals of time, follow through the tasks in the same order. Company management has each task clearly defined, both in terms of content and duration.
3. The operators and tools perform elementary operations that correspond to a single or small number of tasks. These funds are spatially deployed around the different workstations.

Line processes results in a high degree of sequential rigidity favouring the specialisation of the funds.

A technical and economic analysis of the line process may be divided into two stages. First, the optimum line speed is calculated or, more specifically, the pace of production that will avoid the funds ever standing idle. Second, a study is made of the consequences of increasing line operating speed after it has been fully synchronised.

3.1 General conditions for line deployment

Different authors have investigated the general conditions required to guarantee and maintain the line deployment of a given elementary process (Tani 1986: 217-225; Landesmann 1986: 299-309; Gaffard 1994: 72-77; Petrocchi and Zedde 1990). Such reflections have led to the determination of the time lag required to guarantee the continuous activity of the funds. An interval of time closely linked to that of the duration of the different phases (or stages) of the elementary process.

3.1.1 The problem and its generalisation

Take an artisan type production process in which the human work fund plays an outstanding role. For the sake of simplicity, it is assumed that the activity times of the remaining funds coincide fully with those of the labour fund. To all events, interim intervals of inactivity still exist, such as the waiting period after applying dye to a cloth. Let us then look at the hypothetical elementary process shown in figure 23, comprising three stages (tasks or phases) of 12, 4 and 10 hours each. We shall call these d_s, s denoting the order in which the stage appears in the process. Thus, $d_1 = 12$, $d_2 = 4$ y $d_3 = 10$. The total duration of the elementary process is 34 hours ($T = 34$), given that there are two intervals of inactivity, of 6 and 2 hours each.[1]

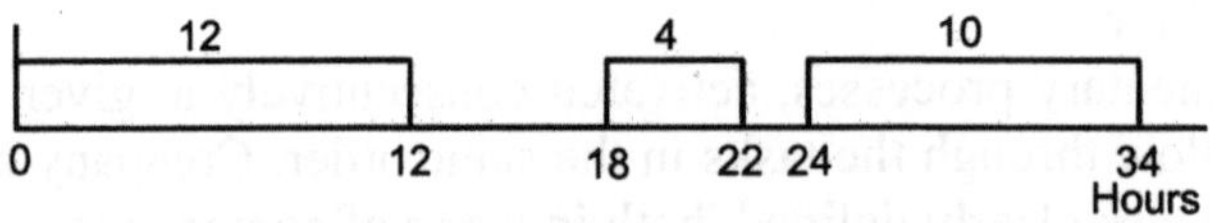

Fig. 23. A hypothetical elementary process

The elementary process in question could have been deployed as per any of the three known patterns: sequential, parallel or line. However, neither of the first two would have avoided the presence of fund idle times, even if they participated in all three phases of the process. Indeed, sequential activation implies the funds remaining inactive during 23.5% (8/34 hours) of the duration of the elementary process (T). On the other hand, a parallel system, meaning that each fund would only serve one elementary process at a time, would only worsen the situation further: the cost of idle time, per unit of time, would be multiplied by the number of funds employed. Therefore, any solution to the problem of inactivity implies the deployment of the elementary process in line. To obtain the maximum benefit from this, it is important to calculate the right time lag between consecutive elementary processes, i.e., that which eliminates any interruption of fund activity. In finding this result, the following two factors should be taken into account:

1. Elementary processes are divided into stages, the duration of which does not necessarily have to coincide, as in the example given.
2. In order to avoid the funds standing idle in the intervals of inactivity between the stages, they should be capable of rendering services alter-

[1] We could equally have examined an elementary process with no intervals of inactivity, but such processes are less commonplace.

nately to the different successive output units. Transferring a fund from a unit of output in process to another can be done at zero cost or, at all events, holding a cost lesser than that of fund inactivity.

In the case in point, in order to avoid interrupting fund activity, the time lag at which the successive processes start is two hours. In effect, this value is the *maximum common divisor* (hereinafter MCD) of the duration of the stages of the elementary process. Thus, the first stage, (d_1) lasts 12 hours (3x2x2), the second, (d_2) 4 hours (2x2) and the third, (d_3), 10 hours (5x2). Consequently, the MCD is 2. This is because it is the largest integer number contained a precise number of times in all the stages of the elementary process considered. The MCD indicates the pace of activation of processes in line, without any dead time which is compatible with the execution of all the stages and the activation of the funds. If this number is called δ, then each δ units of time, the production is equal to one output unit. Or, in other words, for each unit of time, $1/\delta$ output units are obtained. In the case shown above, $\delta=2$, meaning that every 2 hours one product unit is completed.

Time lag δ obliges the funds to work on different elementary processes. They will all *move* from one output unit in process to the next, without any break in the continuity of the services rendered. Moreover, because of the assumption that fund units are unable to operate simultaneously on several elementary processes, a number of units of each fund will be required.[2] The questions that now come to mind are the following:

1. How many elementary processes?
2. How many fund units?

The number of processes simultaneously activated will depend on the ratio of total duration (T) to time lag (δ). For the example, $T/\delta=34/2=17$. Figure 24 illustrates the line activation of the elementary process over time (*x*-axis). As can be seen, between moment t_1 at which the first process starts, and t_2 at which it is completed, up to 17 other elementary processes will be underway. Obviously, this is the number of times the value δ is contained in T.

The number of processes simultaneously activated in a given plant is defined as the *Size of the Line Process*: $S_p=T/\delta$ (adapted from Scazzieri 1993: 32). This variable bears a direct relationship to the length of the elementary process (T) and an inverse relationship to the duration of the time lag (δ).

---

2 Given their complementarity, the assumption is made that the sets of funds (worker and tools) coincide with the work sites or stations. However, this assumption does not need to be rigidly adhered to.

The duration of the intervals of time between such different stages and their duration determines the scale of the line process.

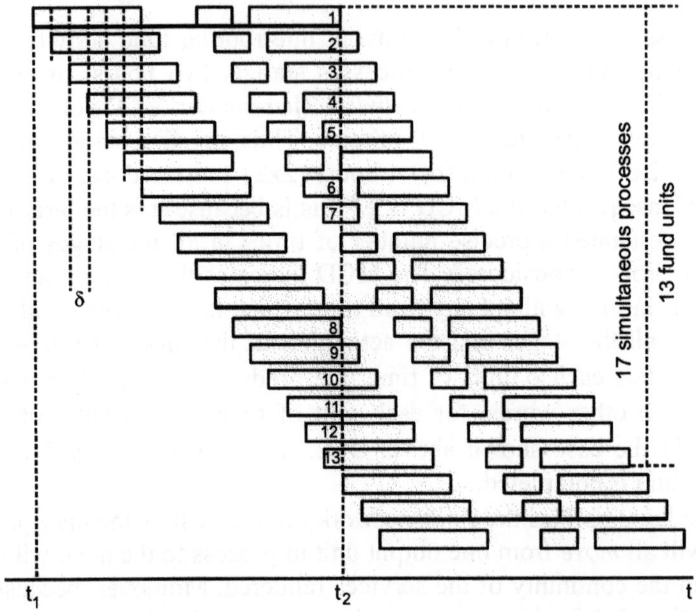

Fig. 24. Values S_p, δ and ϑ in a line process

The number of units of each fund activated at the same time (ϑ) will depend on the number of time segments equal to δ which are contained in the stages of the elementary process. These fragments are denoted v_s. In the example described, the stages last a total of 26 hours (12+4+10). That is, 13 times the value $\delta=2$, a figure which corresponds to the sum of the values of v_s. Indeed $d_1=12=6x2$, $d_2=4=2x2$ and $d_3=10=5x2$, meaning that, given $\delta=2$, $v_1=6$, $v_2=2$ and $v_3=5$. Thus, $\vartheta=\sum_{s=1}^{s=3} v_s$. Consequently, 13 units of the different funds would be required to guarantee the continuity of the line production process. All these funds will operate together and without rest.

The feasibility of the line deployment may be corroborated by seeing what happens when the above-stated conditions are not met. Taking the same elementary process data as above, it is assumed that a δ value of greater than 2 is taken. For example, $\delta=4$. Looking at figure 25 makes it easy to see how imbalances will occur. For instance, the fund that has completed the third stage of the first elementary process will not find the beginning of the third stage emerging from any other process. An option would be to wait 2 hours prior to starting activity with the fourth of the processes. Another would be to connect with the second stage of elementary process 5. That would, however, interfere with the pace of activity of the funds assigned this. Looking carefully at the figure, it is easy to see that a profusion of such difficulties appears.

In figure 25, the size of the line process is not an exact number, nor is the number of funds employed: S_p=34/4=8.5 and 26/4=6.5, respectively. As can also be seen in this figure when the first process is completed, no other starts up immediately thereafter. In effect, the ninth process has already been active for two hours while there are still two hours wanting for the tenth to start. This then shows the inefficiency of taking too great a time lag between consecutive units of an elementary process.

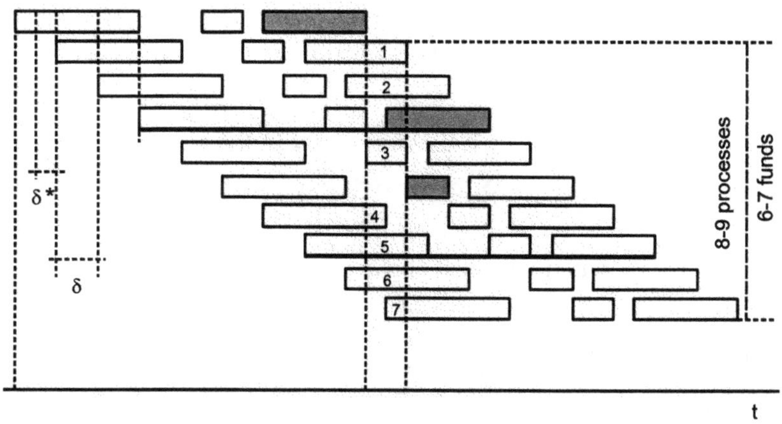

Fig. 25. Inefficient time lag

To conclude with the example suggested, it is the value $\delta^*=2$ which establishes the threshold for the efficiency of the line activation of the elementary process proposed. Values greater than $\delta^*=2$ will cause the different processes to be out of step, inevitably leading to increasing periods of idle time for one or more of the funds. On the contrary, as shown below, lower values will lead to higher output levels giving rise to the mobilisation of an increasing number of funds. For this reason, henceforth the *maximum time lag* for the *optimum* functioning of a given line process, shall be termed δ^*.

It is easy to generalise on all that has been said up to now (Tani 1976: 137 and 1986: 219-222). Let us now take a process that employs diverse fund elements, the jth of which is in process in the following timed stages:

$$\{d_{j1}, d_{j2}, ..., d_{jh_j}\}$$

These stages are separated by lapses of inactivity of any given length. If the jth fund has a single period of activation, that is h_j=1, it could occur that d_{j1} =[0, T]=T, the total duration of the elementary process. Obviously, this may also not be the case: d_{j1}<T, with an interval of inactivity at the start and/or end of the process. To all events, if h_j>1 interim lapses of idle time will appear. Then,

$$\sum_s d_{js} < T$$

The intervals without productive operations will mean that the sum of the periods of activity (s) will be less than the total duration of the elementary process.

If the d_{js} are of commensurable size, there will then be an MCD corresponding to duration δ^* such that,

$$d_{js} = v_{js} \cdot \delta^*$$

with v_{js} being a integer number, for s=1, ..., h_j and j=1, ..., m. In other words, the duration of the d_{js} is v_{js} times duration δ^*. Thus,

$$\sum_s d_{js} = \delta^* \cdot \sum_s v_{js}$$

Moreover, in line with that stated above,

$$T = S_p \cdot \delta$$

If, on the other hand, the elementary process is activated in line with a time lag of $\delta \leq \delta^*$, after a given instant no greater than T, the jth fund will be permanently active, leading to the mobilisation of ϑ_j units. Evidently then,

$$\vartheta_j = \sum_s v_{js}$$

That is, v_{js} is the number of times that time lag δ is contained in the s intervals of operation. For example, remembering the values of the case given above, for $\delta = \delta^* = 2$,

$$d_1 = 12 = 6x2,\ d_2 = 4 = 2x2 \text{ y } d_3 = 10 = 5x2$$

and thus

$$\sum_{s=1}^{s=3} v_{js} = 6+2+5=13.$$

For δ= 1<δ^*, d_1=12x1, d_2=4x1 and d_3=10x1, where

$$\sum_{s=1}^{s=3} v_{js} = 12+4+10=26$$

All of the above then is based on the hypothesis that there is commensurability between the stages of the elementary process (d_{js}). Even though this may not be the case, it is still possible to operate the process in line. Indeed, the ad hoc lengthening of one or more of the stages would permit such commensurability. To all practical effects, this tends to be feasible as it is not considered convenient to reduce the lapses of inactivity of the funds to zero. So there is a certain degree of elasticity, to help cope with unforeseen events that oblige the pace of work of the establishment to vary.

3.1.2 Line deployment and breakdown of the elementary process

The general conditions required to successfully deploy elementary processes in line are also affected by their fragmentability. The line execution of the separate parts of a given elementary process, rather than the whole, could give rise to improved levels of efficiency. In taking an elementary process with stages of different lengths, either including intervals of inactivity or not, the line activation of these stages in different plants could mean having a lower number of processes open simultaneously and, therefore, a potential reduction in the number of funds required. Note that such a reduction would have its repercussions in terms of the volume of production per unit of time.

Let T be denoted as the total duration of the elementary process and T^F (F=1, 2, ..., n) as the duration of its different fragments (stages or phases). Moreover, terms δ and δ^F are the respective values for the optimum deployment. Given that δ is the maximum common divisor of the intervals of activity of the elementary process as a whole, then $\delta^F \geq \delta$. In other words, $\delta^F = \xi^F \cdot \delta$, with ξ^F being a positive integer number. With all of these values:

1. The size of the line for the full deployment of the elementary process as a whole will be given by $S_p = T/\delta$; while, if the different stages were activated separately, the number of processes simultaneously active would be $S_p^F = \dfrac{T^F}{\delta^F}$. As can be find out, $S_p > S^F{}_p$, given that the different T^F are parts of T and $\delta^F \geq \delta$.

2. The number of units of funds for the process as a whole is

$$\vartheta = \frac{1}{\delta}\sum_{s} d_s$$

while for the different fragments it is

$$\vartheta^F = \frac{1}{\delta^F}\sum_{s} d_s^F = \frac{1}{\xi^F \cdot \delta}\sum_{s} d_s^F$$

According to the above, it is shown that the total number of funds employed in the separate deployment of the stages of the process will always be equal to or less than those required for the activation of the undivided process. Now, if it is wished to employ the same number of funds (ϑ), the total number of production lines (or plants) required will be given by the following expression:

$$\vartheta = \frac{1}{\delta}\sum_{F}\left(\frac{1}{\xi^F}\sum_{s} d_s^F\right)$$

In other words, low values of δ^F, with respect to δ, indicate the number of specialised plants required for each stage if the same number of funds is to be occupied. This latter would be the case if the elementary process were deployed as a single block.

3. In both cases, the level of production per unit of time will be the same if the number of plants in which the stages are executed separately is multiplied appropriately. That is, in accordance with the rule given above,

$$\frac{1}{\delta} = \xi^F \cdot \frac{1}{\delta^F}$$

An example of an elementary process is shown in figure 26, with phases (or tasks) of the following duration,

d_1=2, d_2=4, d_3=10 and d_4=20 hours

and with a 6 hour interval of inactivity between d_1 and d_2. Then $T=42$. In addition, it is assumed that the same fund (ϑ_1) performs the first two phases (or tasks), while another (ϑ_2), performs the remaining two.

If the process is activated in one single plant, then $\delta=2$, S_p=21, ϑ_1=3 and ϑ_2=15, with a production level of 1/2 unit per hour. If, however, it is decided to fragment the process, employing two separate plants, then:

1. T^1=12 (d_1=2, d_2=4 with a 6-hour lapse of idle time between stages).
2. T^2=30 (d_1=10 and d_2=20).

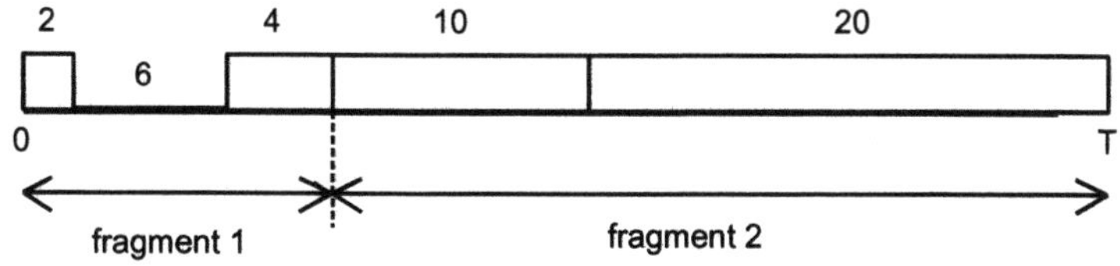

Fig. 26. Line activation and the breakdown of the elementary process

With these figures then,

1. $\delta^1=2$, $S^1{}_p=6$, $\vartheta_1=3$ and the level of production is 1/2 unit per hour.
2. $\delta^2=10$, $S^2{}_p=3$, $\vartheta_2=3$ and the level of production is 1/10 unit per hour.

Given the 15 fund units of the second type, up to 5 plants performing this particular fragment of the elementary process may be activated. In total, breaking down the process has represented the use of six rather than one production plant. With these values, the overall production level is maintained at one half unit per hour.

The fall in the number of processes simultaneously activated (S_p) in the above example, from 21 to 6 and then 3, could result in a manufacturing cost saving for the company, as it facilitates the possibility of benefiting from any partial increases in productivity (that is, in certain tasks or stages). In effect, segmenting a production process at different plants reduces the upheaval caused by any technical or organisational improvement, given that it is easier to readjust execution times and balance out the different fund capacities. In other words, a smaller scale line process is easier to control. Moreover, such spatial dispersion may permit access to cheaper funds and flows. However, it also requires the multiplication of the number of production units and thus all the above benefits may be partially eclipsed by the transport and transaction costs (Tani 1976: 140).

In case of industrial processes with short duration phases, limited indivisibility of funds and low transport costs of in and outflows, their execution could be done across countless establishments, ranging from an extensive region to worldwide. This outsourcing takes advantage of,

1. Lower prices of purchased goods and productive services, given that the small subcontracted firms normally use cheaper and more flexible workers.
2. Reduced uncertainty of the supply of semi-elaborated inputs (delivery times, amounts and quality).
3. The spreading of production amongst several small workshops makes it easier to match production to the changing demand.

It should be stressed that the fragmentability of the production process is an essential requirement for extending the practice of outsourcing though not sufficient in itself. Obviously, many other factors come into play, such as the cost of transport and communications, the ease of incorporating technical innovations limited to a particular stage of the production process, labour policy, and so on (Landesmann and Scazzieri 1996c: 261-270).

3.2 Optimising the timing of the line process

Having broached the issue of optimising the times of the line process following Georgescu-Roegen's proposal, it is possible to go farther. In this sense, it is very interesting to connect the fund-flow model with a set of topics of the management of production processes. Indeed, for the following pages, the analysis will enter the field of operational research, taking into account the practical tools commonly used to balance processes in line.

As it is well known, the question of optimising times in a productive process may be tackled on a minimum of three different levels:

1. The first, and most abstract, is to calculate the optimum time lag (δ^*) between consecutive elementary processes so as to eliminate any fund idle time. This is the type of analysis that has been done above.
2. The second level is a little closer to reality: it investigates the *general* problem of balancing a production line. Given a certain objective of production volume, the analysis tries to determine the appropriate distribution of the different tasks (known their precedence order and times of execution) amongst the diverse workstations, provided that the line is balanced. To achieve that result there are some formal procedures of calculation developed by the production management theory.
3. The third is to establish working rules, incentives and so on, in order to adjust the workstation activity to the production planning decisions taken by the company.

Everyday production management attempts to comply with the level of demand without delays coupled to the goals in production and storage costs. On the contrary, the model developed herein has a limited scope. The analysis is restricted to the second of the levels given above. It is merely a description of the problem of time balancing of assembly lines. Exploring the casuistics of optimisation lines enters within the realms of detailed production management and operational research, i.e. the third of the levels mentioned above (Amen 2001; Baudin 2002: parts A, B and C;

Heizer and Render 1993: chap. 9; Vonderembse and White 1991: chap. 8; Wild 1972: chap. 3, 1989: chap. 15).

As cited in the previous chapter, the assembly line is a particular form of line production system that comprises several workstations deployed alongside each other for the conveyance of the output in process. Assembly lines are widely used for the processing of large quantities of the same product or mass production. In terms of optimisation, such processes present two major problems:

1. The best *sequencing* of the different tasks.
2. The *balancing* or most regular assignation possible of operating times for a preset volume of production.

Here the question of the order of tasks will be ignored[3] and the analysis will be limited to the balancing of the tasks executed by the different workstations.

Although all assembly lines offer their own peculiarities, they are distinguishable by the following main traits (Scholl 1999: 6-19):

1. With respect to the existence or not of a homogeneous work cycle, we can speak about paced (set pace) or unpaced (flexible pace) lines. With regard to the latter, as the workstations present notable differences in worktimes, an inventory of the goods in process will be required.
2. Depending on the number of products manufactured, there are *single-model lines* for assembling slight different versions of the same product with no loss of continuity. If the differences are more notable, a *multi-model line* would be used to assembly the goods in separate batches. This latter case would probably require the line to be stopped momentarily to prepare the changeover. Such breaks are known as *set-up times*.
3. In accordance with the nature of the operating times. Thus there are assembly lines with tasks performed at set intervals while, in other cases, the times may vary in a random but delimited way. It must be remembered that the *launch interval* of the line (or the interval of time required for unloading the parts) is another factor that influences the nature of the operating times. Thus, the feed rate may be set or variable.
4. Depending on the layout of the line. Here the distinction is between lines with the workstations placed one after the other, replicated in parallel, in a straight line, loop or U-shape.
5. With respect to the number of workers employed at each workstation. There could be one single worker, or a group.

3 This question is assumed to be an issue related to engineering.

There is no doubt that there are many other features that characterize several types of assembly lines. Despite such variety, the question of balancing is always basically configured in the same way. In should not be forgotten that the problem of smoothing the work-times has enormous economic consequences for a firm: assembly lines tend to require substantial capital investment giving rise to facilities with a long productive life, it being of a great interest to exploit their productive potential to the full.

The basic goal of any assembly line design is to assign an equal workload to each workstation, that is, to make the intervals of time for the execution of the different respective task/s identical at all workstations.[4] In effect, if the total time required to complete a given phase of a productive process is divided into equal parts for the different workstations involved, the flow of output in process will circulate smoothly and regularly. This will avoid any inefficiency caused by bottlenecks, waiting periods, etc. because one workstation requires more time to complete its work than the rest. Nonetheless, only rarely is perfect synchronisation achieved. More often the work-times are adjusted until the line operates in a way which is considered sufficiently satisfactory. This is equally true of both labour intensive or mechanised facilities.

The cycle time (c) refers to the time available for the workstations to perform their respective tasks. The cycle time comprises both service time and idle time. The service interval may be divided into two components: the *effective* working interval (or transformation work time) and the non-processing time. It is in the period of effective working that the funds are actually processing the output in process, while the latter is the time required to move tooling, load and unload jigs, test the product, convey the output from one workstation to another and so on. It could be considered as auxiliary production time. Look at the hypothetical example shown in figure 27.

The figure shows a phase where diverse funds are separately active at five different workstations.[5] One or more tasks are performed at each. The service times are s_i ($i=1, ..., 5$). These include both the effective work (p_i) and the non-processing (or non-transformation) intervals (π_i), that is,

$$s_i = p_i + \pi_i$$

---

4 As is known, assembly lines often combine workstations with one single worker, with others with a group of workers (multi-task workstation). This analysis does not consider such details.

5 The chart shows the tasks in vertical direction.

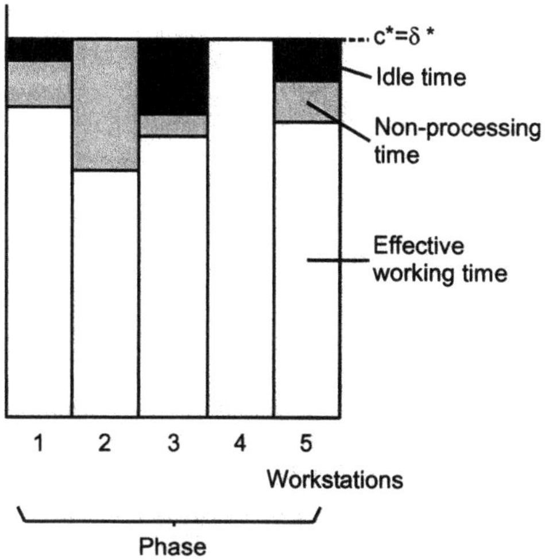

Fig. 27. Tasks and time

As can be seen, the first, third and last workstations present idle time. This idle time is a wasted time. This is the length of time the stations that have completed their work have to wait because other workstations are still active. This is the case of the second and fourth workstations in the example proposed. This will always occur when a workstation has a service time that is shorter than the cycle time which is set by the workstation that has the longest service interval. For this reason, idle time may also be called *balancing delay*: the span of time required for the workstations to achieve a common cycle time (c^*). This value, as can be seen in figure 25, is equivalent to the optimum that indicates the pace of activation of a process in line. In short, $c^*=\delta^*$.

The complete conceptual framework is,

$$c^* = s_i + e_i = p_i + \pi_i + e_i = \delta^*$$

if the delay or idle time is denoted as e_i.

Having clarified the nature of the problem of balancing a line, the analysis may continue after having established the following assumptions, made for the sake of simplicity:

1. The output generated by each elementary process is unique and homogeneous. At the same time, it is obtained at the last instant of its end.
2. The technical features of the flows and/or funds are given.
3. The elementary process comprises one single phase. This permits us to leave aside the problem of connecting interim process stages.

4. The number of tasks in each elementary process is n, to be distributed amongst m workstations. The ratio between workstations and tasks is $m \leq n$.
5. The timing of specific productive operations within the service intervals (s_i) of the workstations is not known. In reality, the analysis only requires knowing the production cycle time (c).
6. The line runs with strict operating times equal to the feed rates. The times required to transfer the output in process between the different workstations has not been taken into account.

Furthermore, to reduce the problem of the heterogeneity of the funds as possible, the analysis has been done considering a manual assembly line where is human work is the most salient input fund. On the one hand,

manual flow lines or assembly lines are characterized by their use of manual labour, for either or both the transfer of items or material along the line, and the performance of the necessary work on the material or items (Wild 1972: 45-46).

On the other, transfer lines have been defined as a

series of automatic manufacturing tools connected by work-transfer devices. Such lines are normally used for metal (or material) cutting and working, but assembly-type transfer lines are currently being developed. Although early transfer lines relied on the manual movement of the material or product, most contemporary lines utilize automatic movement or transfer methods (Wild 1972: 44).

In short, a production system in which the main function of the worker is process control. The issue of balancing automated lines will be attempted in a further section of this chapter.

Figure 28 shows a general scheme for the two-stage resolution of the problem of optimising the times of (manual) production lines.

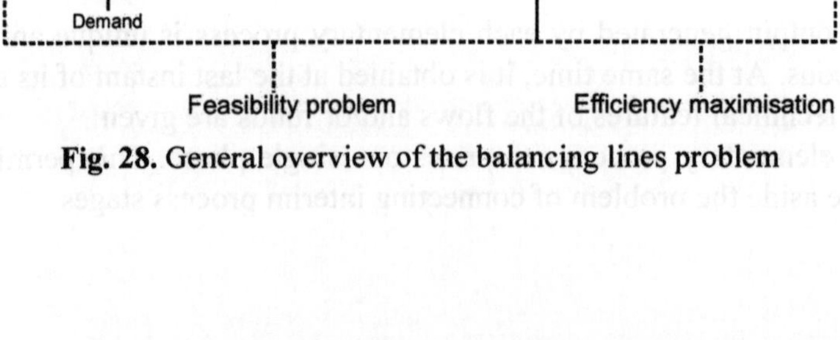

Fig. 28. General overview of the balancing lines problem

A study of the optimisation of an assembly line will start with a graph of the task precedence order. This diagram supposedly reflects a given level of technical and organisational know-how. Look for example at the case presented in figure 29.

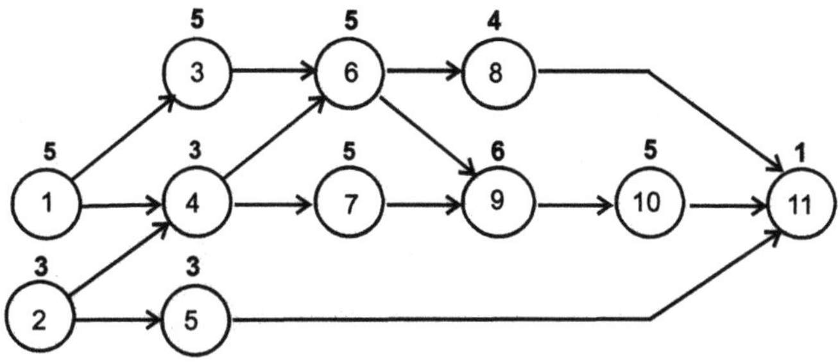

Fig. 29. Example of a task precedence graph

The figures inside the nodes indicate the timing order of the tasks (n=11), while those outside refer to their duration (t_j) given in minutes. The sum of the service times (t_{sum}) is therefore 45 minutes.

Furnished with the information given in the precedence graph, it is possible to consider the *feasibility problem* for the proposed assembly line. Indeed, prior to working on the optimisation of any assembly line, it must demonstrate its viability. The feasibility of assigning a given precedence graph for the tasks (each with its own execution time) to a specific number of workstations (m) is a problem which depends on the time horizon (or total time) available to produce a given volume of output. The last two exogenous figures will determine the cycle time value (c). In the example examined here the goal is established of producing 40 output units during the course of an 8 hour working day. Then the cycle time is c=12 minutes per unit.

The line feasibility means the different tasks will be appropriately distributed amongst the different workstations. Without wishing to delve into the mathematical algorithms that would permit the resolution of the problem, it should be pointed out that, to start with, the procedure would comprise calculating the earliest workstation (E_j) to which a given task (t_j) may be assigned. This is obtained through the expression,[6]

---

[6] The symbol $\lceil \cdot \rceil$ indicates the closest integer number greater than or equal to (.). In other words, E_j is the result of rounding the expression up to the nearest integer number, unless (.) is already an integer number, in which case, it will be used as such.

$$E_j := \left\lceil \frac{t_j + \sum_{h \in P_j^*} t_h}{c} \right\rceil \quad \text{for } j=1, \ldots, n$$

in which P_j^* is the set formed by the immediate transitive tasks that precede the j task considered.

The same calculation is made to ascertain the last workstation (L_j) to which a given task may be assigned. The expression is the following:

$$L_j := m + 1 - \left\lceil \frac{t_j + \sum_{h \in F_j^*} t_h}{c} \right\rceil \quad \text{for } j=1, \ldots, n$$

in which F_j^* is the set formed by the immediate transitive tasks that follow task j, and m the desirable number of workstations.

The t_j, E_j and L_j values are shown in the table below.

J	1	2	3	4	5	6	7	8	9	10	11
t_j	5	3	5	3	3	5	5	4	6	5	1
E_j	1	1	1	1	1	2	2	3	3	4	4
L_j	2	2	3	3	4	4	4	4	4	4	4

The top row indicates the task to which each value corresponds. Additionally, the table indicates the earliest and last workstations to which a given task may be assigned, bearing in mind the c value (no workstation may perform a task that takes longer than c) and the conditions $E_j \geq 1$ and $L_j \leq m$.

With these values, and assuming that $m=4$, the tasks are assigned to the different workstations. This means calculating the intervals $SI_j=[E_j, L_j]$ referred to the sets of tasks which may *potentially* be assigned to each workstation. These sets are called B_k and are constructed as per the expression: $B_k = \{j | k \in SI_J\}$. In the specific case in hand, the values of the last two rows of the above table may be rewritten as follows,

J	1	2	3	4	5	6	7	8	9	10	11
	1	1	1	1	1						
SI_J	2	2	2	2	2	2	2				
			3	3	3	3	3	3	3		
					4	4	4	4	4	4	4

Therefore, the B_k sets are the following:

$$B_1=\{1, 2, 3, 4, 5\}$$

B_2={1, 2, 3, 4, 5, 6, 7}

B_3={3, 4, 5, 6, 7, 8, 9}

B_4={5, 6, 7, 8, 9, 10, 11}

Then tasks may be definitively assigned to the different workstations. To this end, the following observations must be taken into consideration,

1. The sum of the times of the different tasks may not exceed the duration of the cycle.
2. There must be a direct link, as per the precedence graph shown, between the tasks performed on one single station.
3. The B_k sets should be read strictly in the order in which their values are shown.

Finally, following the above criteria of selection, we obtain,[7]

B_1={**1** (5), **2** (3), 3 (5), **4** (3), 5 (3)}

B_2={~~1~~(5), ~~2~~ (3), **3** (5), 4 (3), 5 (3), **6** (5), 7 (5)}

B_3={~~3~~ (5), 4 (3), **5** (3), ~~6~~ (5), **7** (5), **8** (4), 9 (6)}

B_4={~~5~~ (3), ~~6~~ (5), ~~7~~ (5), ~~8~~ (4), **9** (6), **10** (5), **11** (1)}

As can be seen, the tasks grouped at each of the four workstations are shown in bold, while the numbers crossed out indicate the tasks already assigned.

The same procedure of assigning tasks to workstations may be used following certain formal rules. The most simple suggests the definition of the dichotomic variables x_{jk}, which will show a value of 1 if task j is assigned to workstation k, or 0 in any other case for $j=1, ..., n$ and $k \in SI_j$ (Scholl 1999: 29).

The following restrictions determine the feasibility, or otherwise, of a balanced line for any combination (m, c):

1. $\sum_{x \in SI_j} x_{jk} = 1$ for $j=1, ..., n$.
2. $x_{jk} \in \{0, 1\}$ for $j=1,2, ...,n$ and $k \in SI_j$.
3. $\sum_{j \in B_k} t_j \cdot x_{jk} \leq c$ for $k=1, .., m$.

7 Each task is shown with its respective duration (figure shown in brackets). This is to facilitate an understanding of the procedure by which it was resolved.

4. $\sum_{k \in SI_h} k \cdot x_{hk} \leq \sum_{k \in SI_j} k \cdot x_{jk}$ for $(h, j) \in A$ and $k \in SI_j$.

Restrictions 1 and 2 ensure that each task is assigned to only one workstation within the interval SI_j. On the other hand, the restriction 3 establishes that the duration of the cycle is not exceeded by the activity time of any of the workstations. Finally, the restriction 4 guarantees that no task is assigned to a station that comes before another which is performing an earlier task.

The example shown is a so-called *Simple Assembly Line Balancing Problems* (SALBP) In general, the balancing problem to be undertaken is far more complex. For this reason, there is an increasing need to build very sophisticated mathematical algorithms when attempting to resolve these (Scholl 1999: 30-42). Moreover, it should be stressed that, at times, the problem of balancing is tackled with line programming methods. However, such models require lengthy calculation and hence the development of heuristic procedures to resolve them (Amen 2001). This is beyond the scope of this book.

Figure 30 shows the final result achieved: the assignation of the tasks to be executed to the different workstations (m=4 in this case).

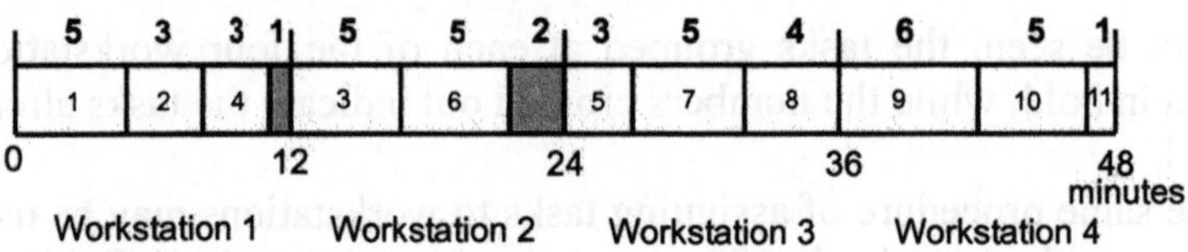

Fig. 30. Elementary process and the workstation task allocation

Despite having encountered the best possible distribution of the tasks, workstations numbers one and two still present intervals (the shaded areas) of inactivity (e_i, or delay time). Therefore, the real capacity of the line (T=c·m) is 48 minutes. Thus,

$$T = c \cdot m \geq \sum_i s_i$$

Having established the basic elements of the line, the following concepts may now be defined:

1. Efficiency (time) of the line (E). It is the ratio of total task execution time to the real time it takes one output unit to run the whole length of the line. That is,

$$E = \frac{t_{sum}}{T} = \frac{t_{sum}}{c \cdot m}$$

The adding of idle times for balancing lowers the efficiency of the line. When the fit is one hundred percent, then E=1. In the proposed case, E=45/48=81.81%. Or, in other words, productive activity occupies 81.81% of total line operating time.

2. The interval of time lost once the line is balanced. This is the time lapse separating T and t_{sum}. In the example given,

$$BD = T - t_{sum} = c \cdot m - t_{sum} = 3 \text{ minutes.}$$

Of great interest is the *Balance delay ratio* defined as BR=1-E.

3. The indicator,

$$SX = \sqrt{\sum_{k=1}^{m} [c - t(S_k)]^2}$$

measures the degree of homogeneity of the distribution of work between the different stations (*smoothness index*). In this expression, $t(S_k)$ is the total service time consumed by each station (k=1, ..., m). In the case in hand,

$$SX = \sqrt{1^2 + 2^2 + 0^2 + 0^2} = 2.23$$

The closer the resulting figure is to 0, the better balanced the work load is at the different stations.

With all the above information, it is plain that the main problem for optimising an assembly line is *maximising efficiency* (E). Given that E depends on a non-lineal term, i.e., the product of the cycle and the number of stations ($c \cdot m$), this goal may be divided into three particular cases:

1. To maximise the number of stations m, for a given c. This is a goal which tends to be attempted when working with a new assembly line to be installed and the estimated output demand is known.
2. To minimise the cycle, for a given m. This second case tends to occur when the object is to maximise the rate of production of an existing line. It is the kind of problem that frequently occurs in what is known as process re-engineering.
3. To achieve the highest possible level of line efficiency, seeking the best possible combination of m and c. This is the objective examined in the analysis below.

An attempt to achieve the highest possible level of activity during total line operating time is equivalent to minimising line capacity (T), or minimising BD (or BR).

To begin with, it has to consider the existence of a hypothetical minimum number of workstations. Due to the division of the work, the lower limit to the number of stations (m_{min}) cannot be $m_{min}=1$, with $c=t_{sum}$ and $E=1$. Although it might be an ideal solution, it must be acknowledged that the activity times shown in the task precedence graph already include, in terms of productivity, the benefits drawn from the division of the work. To all events, it must be pointed out that, owing to the condition $T=c \cdot m \geq t_{sum}$, the lower limit to the number of workstations would always be given by the ratio,

$$\frac{t_{sum}}{c_{\max}}$$

obviously rounded up to the nearest integer number, or in other words, given that t_{sum} is a constant, the minimum number of workstations will be associated with the longest lapse in the work cycle.

Optionally, there may also be a top limit (m_{max}), related to the limitation of space in the plant or availability of labour. It is therefore a limit set by factors external to the issues of the assembly line itself.

With regard to the cycle, the lower limit (c_{min}) is derived from the condition $c \geq t_{max}$, where t_{max} is the duration of the longest task. In practical terms, this limit is influenced by the possible maximum volume of sales per unit of time (or the capacity of the market to absorb the output). Nonetheless, a lower limit for c may always be found by applying the following ratio,

$$\frac{t_{sum}}{m_{\max}}.$$

In other words, given that t_{sum} is a constant, the fastest pace of work corresponds to the highest number of workstations. In terms of the upper limit (c_{max}), this may be due to the existence of a desired rate of production.

Summarising the above limitations, the problem of optimisation is defined by two intervals [m_{min}, m_{max}] and [c_{min}, c_{max}]. Moreover, to be consistent with the limitations given above, it is assumed that for each m or c value, there is at least one value for the other variable and thus $m_{min} \cdot c_{max} \geq t_{sum}$ and $m_{max} \cdot c_{min} \geq t_{sum}$.

In line with the precedence graph proposed, given that the maximum cycle time is $c_{max}=25$ minutes per unit, where $t_{sum}=45$ minutes, at least $m_{min}=2$ stations would be required. If there were no limits to sales, the lower limit of the cycle would be given by $c_{min}=t_{max}=6$ minutes per unit. Finally, if it is assumed that for questions of space available, under $m_{max}=9$ workstations can be installed, the problem is circumscribed by $m \in [2, 9]$ and $c \in [6, 25]$.

Figure 31 shows all the possible combinations of c and m. It shows the upper and lower limits of c and m. Moreover it shows a set of points (joined by segments merely for the sake of clarity), which correspond to the successive combinations of c and m (always formed by positive integer numbers).

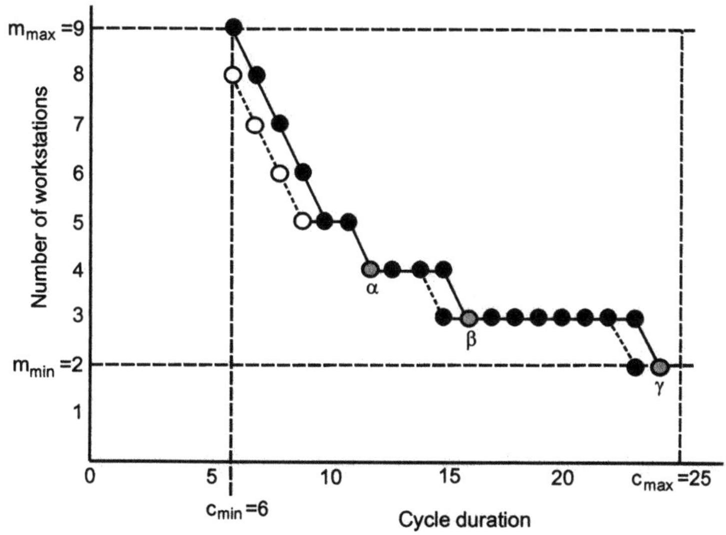

Fig. 31. Combinations of c and m

The region above the curve contains other combinations that meet the essential condition $c \cdot m \geq t_{sum}$. Despite being feasible, these combinations are not optimum. The points on the more or less convex pathway shown in figure 31 are the only ones to be so. The denser points indicate the minimum number of stations for all possible c values, but only the highlighted points are related to a maximum level of efficiency. That is, combinations $\alpha=(12, 4)$, $\beta=(16, 3)$ and $\gamma=(24, 2)$, attaining $E=93.75\%$. As can be seen, value $m=4$ was proposed to illustrate the task assignation contained in the initial precedence graph. On the other hand, the hollow points show impossible line balance pairs (c, m), even though they meet the feasibility requirement $c \cdot m \geq t_{sum}$.

Although there are mathematical procedures with which to find the points shown on the above figure (Scholl 1999: 65-6), in very simple cases such as the one in hand, it is sufficient to examine all the possible combinations (c, m) formed by integer numbers. For each of them, the resulting product T is calculated and consequently an E value is associated. The points we are seeking are those corresponding to the lower E values.[8]

Now, if the goal established is to maximise the production of the assembly line, after having found the optimum (c, m) combinations, the α, β or γ pairs that generates the greatest output per unit of time ($60/c$, being c is measured in minutes) should be chosen. The response is clear: α is chosen, given that it produces 5 output units per hour, as opposed to the 3.75 units of β or the 2.5 units of point γ. Therefore, the same level of (time) efficiency will give rise to different levels of production per unit of time. The reason is the degree of division of the work: lower c values imply greater number of workstations with fewer tasks assigned to each. Production will increase as c falls, as it will be equal to $60/c$ or $60/\delta$. See figure 32.

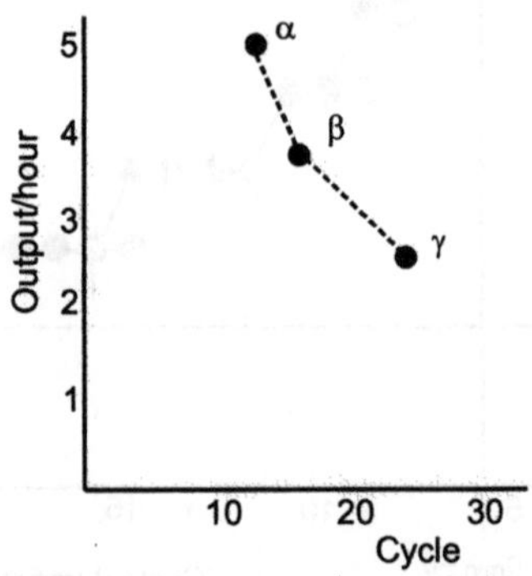

Fig. 32. Level of production and optimum cycle

With the data obtained, production cycle α will take 8 hours to complete 40 output units, while β takes 10.67 hours and lastly γ, 16 hours. The 8-hour maximum limit established demands a rate of production of 5 units per hour.

To aid the understanding of the nature of the optimisation exercise, a figure could be constructed to show the relationship of all the possible cycle variables $c \in [6, 25]$ to the balance delay ratio (BR). Or similarly to calculate the values of $BR=1-E$ for all the different combinations (c, m) situ-

[8] With respect to the above figure, it must moreover be pointed out that the interval [m_{min}, m_{max}] contains far fewer values than [c_{min}, c_{max}], given that the cycle variable is continuous, while the number of stations is quantified by whole numbers alone. Thus, for given m values, there may be a whole interval of rates of activity, all of which are feasible combinations.

ated between the upper and lower limits. After having traced all the points, those with the lowest *BR* values are the best. See figure 33.

In this figure, the discrete points of the resulting function have also been joined by straight lines for the sake of clarity. The pathway of the *BR* values may be explained as follows: while the cycle increases so does the time consumed for balancing, but only until the number of stations required is reduced. Such a reduction would lead to a drastic fall in *BR* values, until reaching a local minimum. In the case in hand, the *BR* value is *BR*=6.25%. On the other hand, the hollow points on the figure indicate those combinations that are not feasible. Finally, the figure also includes the number of workstations operative for the different cycle levels.

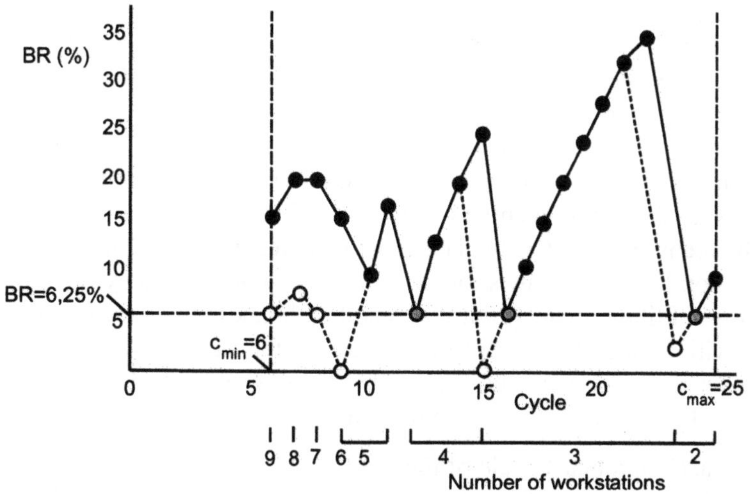

Fig. 33. Relationship between the line balance delay ratio and cycle

To sum up, on confirming that a task precedence graph was compatible with a given number of workstations and work cycle, the analysis went on to investigate whether or not the result achieved also corresponded to the highest level of efficiency and to the highest production volume per unit of time. Obviously, it must be pointed out that the optimisation of an assembly line requires far more complex models than those given on these pages.

3.2.1 Balancing automated lines

Balancing automated lines is a far simpler problem because of the lower number of stations and the more clear-cut tasks usually performed. To start with, the optimisation of the productive circuit, comprising machines capable of executing different elementary operations on parts that move in synch, means that the work times of the stations must be equal. Thus, if t^* is the maximum time required by the longest operation and t_i is the time needed by another station i, the variable degree of saturation (g) of the line may be defined as the ratio:

$$g = \frac{\sum_{1}^{m} \frac{t_i}{t^*}}{m}$$

m being the number of workstations.

In effect, if all the workstation times are equal, then g=1 and, consequently, no time would be lost for delays. Due to deviation, the t_1/t^* ratios are lower than 1, and the sum will therefore always be lower than the value of the divisor. Thus, indicator g will fall below one. The more marked the timing imbalance, the more it will do so.

Despite the above, it must not be forgotten that here we are dealing with fully automated lines. Therefore, we should be confident that the engineers that design them will eventually succeed in balancing the working times of the different stations (g=1). The most troublesome issue concerning such production lines is the question of *reliability*: given that they are integrated systems, a failure at any point will lead to a break in the pace of work. This upset will damage the output in process. At the same time, it must not be forgotten that restarting the line will require a great deal of time and resources. To tackle all that, it should consider:

1. c is the (theoretical) cycle of the line.
2. F is the frequency of breakdowns, power cuts, and other problems that stop the line.
3. T_F is the time associated with inactivity (measured in terms of production cycles).

The mean activity time will be given by the expression,

$$c+F{\cdot}T_F$$

while the pace of production is equal to 60/c (if all time variables are measured in minutes). Therefore, the efficiency (E) of the line will be,

$$E = \frac{c}{c + F \cdot T_F}$$

that is, the ratio of the theoretical and real duration of the production cycle.

Although breakdowns also occur on manual lines, it is assumed that the majority will be resolved by the workers themselves. For example, if a part does not fit or is damaged, the operator is expected to replace it immediately. The same is true with any damaged tooling. This is not the case with automated lines. In addition, the probability of a failure occurring on any specific automated line is given by,

$$F = 1 - \prod_{1}^{n}(1 - f_i)$$

If the f_i values are all assumed to be equal, then the expression can be simplified,

$$F = 1 - (1 - f)^m$$

m being the number of workstations. This expression highlights the fact that the frequency with which breakdowns occur will rise as the number of workstations or cells increases. Thus, the longer a line is, the less reliable it will become.

3.2.2 The line process and the specialisation of funds

After briefly having explained the problems of optimising the times of line processes, it is essential now to return to the question of the specialisation of the fund elements. Although line production is not necessarily coupled to the technical division of the work, there is no doubt that, as opposed to the other forms of deploying an elementary process, it facilitates it greatly. Funds could be specialised in sequential or parallel production as well, but it is only line deployment that guarantees that they are fully occupied all the time.

To start with, it should be said that Adam Smith already warned of the difficulties of specialisation in production processes based on the seasonal growth of plants and animals. Rather, in an artisan process, the work may be fragmented to suit the different skills of the workers in the different tasks and/or with different tools. Nonetheless, strict specialisation would not be interesting as it would bring with it long intervals of inactivity. A greater skill at a given operation will only be used to improve its execution. For instance, final retouching of an output unit destined to a preferential client is a reason which justifies the higher quality of the task performed. On the contrary, in a line process, work could be unskilled, given that the worker is strictly limited to performing those tasks to which she/he is assigned. Workers will only carry out a given number of simple operations, their scope of action in the production process being seriously delimited. While the role of the worker in sequential production is clearly visible, both in the planning of the activity and in its execution, in a line process, the former is absent and the latter is fragmented and restricted.

The division of labour leads to the progressive simplification of the different production operations. At the same time, these operations are distributed amongst an increasing number of workers, given that the tasks assigned to each will become more and more, narrow. Within the conceptual framework developed here, the elementary process comprises phases which can be broken down into tasks. This hierarchy allows us to distinguish, albeit somewhat basically, two levels of the division of labour:

1. Each fund may execute *one* of the phases comprised in the elementary process in full.
2. Each fund may perform a *fragment* of the phases of the elementary process. In this case, the tasks that result from the fragmentation of the elementary process must last the same time and, moreover, be synchronised with the time lag between consecutive processes (δ), or a multiple of these.

In both the above cases it is obviously assumed that the funds will act without rest on the immediately consecutive processes.

Whether it be complete phases or tasks, it is highly probable that the execution time will be reduced through less time lost on operations, such as changing tooling or moving materials, shorter learning curves for the operators, and so on. Thus, without making any major changes to the production process, its scale ($S_p=T/\delta$) will increase, as will the level of production ($1/\delta$). This is because the specialisation of the funds will have meant shorter cycles of activity, thus permitting a reduction of the initial δ value, as long as none of the workstations is held up. Nevertheless, it must be pointed out that the number of funds, workers or machines, will likewise have increased too. With more workstations needed, each with smaller fragments of the elementary process, more operators will have to be hired and more tooling and machinery purchased. To all events, while the number of funds increases, with respect to the initial value of δ, specialisation will lead to successive improvements in performance, equivalent to the further reduction of δ and thus to the duration of the elementary process.[9]

---

[9] Despite everything stated above, specialisation should not be mistaken for meaning the permanent activation of the funds. Thus, it is possible to conceive of a case in which the funds experience no idle time, yet are not specialised (Tani 1986: 223). This would mean each fund executing *all* of the phases of the elementary process, acting on different output units, activated with the appropriate time lag which would be associated to the commensurability of the different phases of activity and inactivity. It should be pointed out that, in this

3.3 The scale of the line process

Line activation is feasible for any production process that can be broken down into stages and will, moreover, accept the adjustment of its internal times. However, there are reasons beyond the technical sphere which might make line deployment inadvisable. In effect, as the factory system considerably increases the output volume per unit of time, what emerges as a major obstacle is the capacity the market has to absorb the increased supply. While sequential manufacturing will generate one product unit every T hours, line production will produce T/δ, where δ is a fraction of T. Thus it is indisputable that the pressure of demand is an essential requirement for the use of line deployment. However, the Classical school economists had already discovered, the most outstanding feature of the *Factory System* is not only its ability to increase levels of production, but also the fact that at the same time it will reduce unit cost. Thus, the potential demand also expands. This increase calls for the implementation of faster production techniques: the number of products per unit of time increases. Consequently, one embarks upon a cycle of reciprocal interaction, fed, of course, by the spirit of competition.[10]

The above reflections underscore the fact that, amongst the heuristic aims of the model developed here a detailed analysis of the concept of the scale of production is of prime importance. It is a concept that gains theoretical relevance as the productive potential of the line process also grows.

The level of operation or *Scale of a production process* is defined *as the amount of output generated per unit of time*. Although such a definition may, at first sight, appear excessively simple, the analysis of the factors that determine the scale of a process, and the consequences deriving from its change, is of great intricacy. In effect, the scale at which a process operates at any given moment of time corresponds to a multifarious set of technical and economic reasons which, in general, will vary over time. Then, given that all production processes are of complex structure, any at-

case, the element would not act on all of the phases of the *same* elementary process as, if it were to do so, it would not be a line process. Therefore, each fund unit operates on processes which *do not* appear immediately consecutive. However, this is a purely theoretical case. If attempted it would encounter problems related to coordination and to the layout of the different workstations. Moreover, such activation would not allow any benefit to be obtained from the improved skills derived from the increasing specialisation of the workforce.

10 Obviously, things are not as simple as this. Any increase in spending power leads to a change in the make up of demand. Thus, the feedback mechanism referred to is merely suggestive of the general dynamics.

tempt to unravel the most significant aspects related to a change of scale is also a difficult exercise. All the approaches so far made have tended to be incomplete.

In the above definition of scale it has been assumed that:

1. The plant considered executes one single type of elementary process with a single output. If the output is multiple then it is assumed that the proportions of the different sub-products will not vary with an increase of scale.
2. The production equipment and procedures are all defined. The techniques used will not change significantly due to the increase in output volume per unit of time.

It is clear that both the above assumptions are very restrictive. However, the history of economic thought reveals that the theoretical discussions regarding the concept of scale have been muddled, and the empirical assessment of the effect of the change of scale on unit cost has proven arduous.[11] Although the explicit review of such issues is not the object here, the difficulties encountered justify proceeding with caution. Establishing the above restrictive assumptions therefore represents a promising start as it obliges a slow pace. At the same time, this will help to clarify the concept of the scale of production.

With respect to the first of the two assumptions, it must be acknowledged that a change in scale will normally lead to a physical modification of the product. However, the functions tend to remain the same without, of course, forgetting about basic industrial products that, whatever the scale, will maintain a high degree of homogeneity. In itself, this is reason enough to comparatively test the effects of different scales on the same production process. That is, the reality to which the economic analysis is referred remains, for the most part, acceptably similar.

The second assumption is a far harder condition to comply with. Any attempt to use the *same* production techniques on very different scales will be sure to encounter great resistance. In fact up-scaling is only feasible when the innovations that will make it possible are available. It is a known fact that, once the operating features of a process are sufficiently well known, engineers recommend only modest increases in the scale of production. The presence of new techniques in all changes of scale makes prudence advisable. One has to act with care to avoid unpleasant surprises

[11] See, for example, the discussion between Chamberlain (1948, 1949) and Hahn and McLeod (1949), as well as the contribution of Lerner (1949). With respect to the issue of empirical assessment, see Pratten (1971) and Smith (1971). All the above bibliography, and much other, is tackled in Gold (1981).

during the start up of a new and larger production plant.[12] Now, with the exception of highly automated processes, the pace of activity may increase without the productive infrastructure having to undergo any significant change. It is possible to analyse a change in scale within the same technical framework. In effect, the very fact of deploying an elementary process, previously performed sequentially, as a line process, gives rise to an increase in the scale of production, without it being absolutely essential to change the productive operations contained therein. This then explains why, from the very outset, the Factory System was seen as more of an organisational than a technical innovation. It is, however, true that later developments have involved both organisational and technical change.

In short, the two above-mentioned assumptions are extremely demanding, but do not prevent us from performing an analysis of the scale of production processes. Indeed, there are many processes for which the output will suffer little change when the scale of manufacturing is varied and, at the same time, the concept of scale refers to a set of factors that is far more varied than has generally been believed to date. Thus, the term *scale* has several different definitions. A clarification of the term is pursued below.

Alongside the definition of scale, one must also clarify the meaning of the concept of *plant production capacity,* or production unit: the output per unit of time (hour) multiplied by the number of hours of activity per annum (H^T). Such effective capacity must be compared to the maximum: that which would correspond to the plant operating 8,760 hours per annum. Nonetheless, it must be pointed out that such comparisons are always controversial, as the rate of work may vary. Thus the issue of maximum potential capacity is always a moot point.

The definition of plant capacity is simple, but somewhat misleading. In effect, to start with, one should not use the term *size* when referring to a plant. Its physical connotations (space, to be more precise) do not sit at all well with the strictly economic meaning of the concept. Secondly, neither is the option of referring to the scale or production capacity as instant, without considering a sufficiently broad interval, completely satisfactory either. The capacity of a production plant, as opposed to the case of a physical receptacle, cannot be considered without including the passage of time (Georgescu-Roegen 1969: 527). Thus, it is not at all surprising that a

[12] This does not mean that different scales form part of a continuum. In effect, as the number of techniques available at any given moment in time is limited, the scale variable will present leaps or discontinuity. The argument given does not attempt to invalidate the properties of the range of techniques (or scales) available. It merely underscores the fact that, whatever the scale, any increase in it should be progressive.

single production facility may possess many different production capacities (de Leeuw 1962). Changes in the work schedule (shifts per day and number of days worked per annum), alongside changes in performance (which affect scale), separate real and potential production for the same period of time (normally a year).

Returning now to the main thrust of the presented argument, the following classification of the changes of scale is suggested:

1. Increasing the volume of production per unit of time merely by deploying an elementary process in line. This is achieved by lessening the idle times of the funds, and the improved performance obtained with specialised workers and machines.
2. Higher productivity may also derive from the rationalisation of production operations as the result of performing a time and motion study, alongside the implementation of a mechanical transport system to convey the output in process from one workstation to another, that is, a combination of Taylorist and Fordist practices.
3. Finally, increasing the division of labour to the extent that workers may be replaced by machines. In this sense, by mechanising these simplified tasks, the firm, over time, will be able to automate the production. In this case, it is clear that an increase in production per unit of time will require greater technical change (Corsi 2005).

As can be seen at a glance, the above classification coincides with the main ways in which line processes, the form of productive activity most susceptible to all kinds of innovation aimed at increasing the scale of the process, have been developed. Evidently each (sub) sector has its own particular technical-economic history which will account for both its technical dynamic and the evolution of its entrepreneurial configuration. Only a case-by-case study will allow us to elucidate the specific role played by up-scaling and its effects (Chandler 1990; Utterback 1994). However, that would be going far beyond the scope of this book.

3.3.1 Expanding the scale of production

As is remembered, the optimum values for the cycle (c or δ) and, by extension, for the number of workstations, have been worked out above. Moreover, it has been demonstrated how values higher than these will lead to a worsening of the efficiency of a line process. Now, the aim is to examine how certain essential features of the tasks and their execution could be modified in order to expand the scale of process in line.

Generically, the organisation of a process in line will run up against three main problems:

1. Identifying the operations required.
2. Defining the capacity of both men and machines.
3. Calculating the optimum flows of materials and output in process.

Having resolved these basic issues, if a higher production goal is established in terms of output per unit of time, the process will have to be recomposed (Gibson and Mahmoud 1990). This means,

1. The increased parcelling of the work process. In doing this, idle time may be reduced and yield increased. That could be attained by hiring low skilled workers who do not require any great outlay in training or long periods of time before they are fully integrated into the process.
2. The use of greater capacity and more reliable fixed funds, leading to a higher degree of process automation.
3. The implementation of control and supervision mechanisms.

The exclusive right of company management to define tasks allows the targets of performance to be transformed into patterns of conduct. Automation and standardisation also contribute to a faster pace of work, in this case normally justified by objective technical reasons. However, automation will not affect all the tasks in a process in the same way. At the same time, automation will be implemented on a trial and error method. Both circumstances make it essential to secure the collaboration of the workers in achieving the corporate goals of profitability.

Prior to proceeding with the main subject that concerns us here, it should be pointed out that an essential aspect of the organisation of any productive process is the assigning of the different tasks of the elementary process to the different funds available. This will obviously be done according to the degree of specialisation. Formally, each fund U^J is defined in an n-dimensional space C of its capacities with respect to a vector, the components of which are measurable (Landesmann and Scazzieri 1996b: 197ff.):

$$U^J = \{C^J{}_i\}, \text{ where } C^J{}_i \in \Re^n$$

Many capacities possess a clear quantitative form of expression (speed, memory, or whatever). Others have a more qualitative nature, but they may be quantified in following certain generally accepted conventions. To all events, such capacities may be compared within a cardinal space. However, skill results from the sum of capacities and therefore it is a multidimensional concept. To assess it with respect to a given task, it is necessary to weigh the constituent capabilities to provide a single indicator. Despite this, given that the criteria for any such weighting may be highly diverse, any ranking of the skills of the funds with respect to a given task will never be robust.

With the criterion given above and the assignation of the working times to the funds, a *programme of tasks* may be designed, i.e., an application between the space of fund capabilities and the space of tasks belonging to a specific process. Any programme will be complete if all the tasks to be performed have been distributed to the funds available. Thus, any programme is the result of the progressive adjustment of the capabilities of the funds to the technical and time requirements of the process.

Given that the funds may be identified by their different working capacities, any production process is the application of a theoretical set of these to the set of tasks required for a specific production process (Landesmann and Scazzieri 1996a: 198). The different programmes specifying the tasks may probably be ordered according to a criterion of technical efficiency (Landesmann 1986: 289).

Having established the assignation of tasks, the activity to be engaged in by a given human work fund unit will depend on the following three factors:

1. The length of the working day.
2. The length of the effective working time (for each shift).
3. The pace or intensity at which the tasks are executed, directly determined by the speed of the machinery and the systems that convey the output in process and, autonomously, by the accumulated experience of the workers (or learning curves).

While the first of these three variables has fallen steadily since the end of the 19th century, the others two have increased. This has occurred as the result of the fragmentation, simplification and control of tasks, changes to the methods of production (that have enabled the lag between consecutive line processes to be shortened) and, of course, the improved capacities of the machinery.

To sum up, the increasing specialisation of work has influenced the nature of the tasks in the following three ways (adapted from Ippolito 1977: 471):

1. In the short term, the average number of tasks performed by each worker has gradually fallen, although, with respect to the whole of an individual's working life, it may have increased with the corporate practice of internal rotation.
2. Tasks have evolved towards a lower diversity of elementary operations.

3. An attempt has been made to shorten elementary operations, i.e., redefining micro movements to increase speed and eliminate any remaining shreds of idle time.

In a line process, the duration and complexity of tasks, the learning process and the pace of work are all closely related. However, the specific configuration of this is intricate. Generically, the first two variables (duration and complexity) are inversely related to the ease of learning and the pace of work. To all events, a great variety of empirical alternatives are possible.

A reduction in the number of elementary operations to be performed facilitates the swift learning of the procedures and tools to be used.[13] Such familiarity with the work in hand is achieved quickly, even for workers with very little prior skills training. This is a phenomenon closely associated with the loss of skill of the specialised worker, particularly as the degree of fragmentation and automation of the line production process grows. Indeed, if skill is understood as an in-depth knowledge of the variety and potentials of the materials, tools and procedures of a given activity, along with the capacity to successfully resolve any unforeseen problems that may arise, it is clear that a cook has a far greater skill base than does a worker whose task consists of feeding and supervising an automatic machine. Only the fascination for all mechanical or automatic devices could account for such workers being commonly attributed with a great skill bigger than, say, a gardener. In any case the level of qualification, in line with the above definition, does not have to coincide with an individual's level of formal education or training.

Since the execution of a given task improves as the worker gains experience in handling the fixed funds required, it should be added that learning by doing is the method fundamentally used for resolving the problems that may occur with the fixed funds. The problems which are referred to are the ability to perform the initial diagnosis of the origin of a fault, the initiative to solve minor problems as they occur, and the development of possible improvements to the procedures or tools. In general, the problems with respect to the functioning of the fixed funds may be classified into two main categories:

1. The unit does not operate to design specification.

---

[13] It is essential to point out that the type of know-how mentioned here bears absolutely no relationship to superior intellectual skills (creativity, intuition), but does relate to the ingraining of reflex behaviour. The impoverishment of human skills associated with the line process was already being criticised at the beginning of the 19th century (Wing 1967).

2. The operator is not satisfied with the performance of the fund, even though it does operate to design specification.

Such problems may,

1. Have been known of beforehand, i.e., prior to the definitive installation of the fixed fund, but were not duly taken into account (for several reasons including the breakdown in communications between the shop floor and designers of the capital equipment).
2. Have appeared unexpectedly and were either detected directly by the operator of the fixed fund in question or by the operators of other phases of the production process, often downstream.

The two criteria may be combined as shown in table 3 (adapted from Hippel and Tyre 1995). This table gives an overview of the causes of the operating difficulties that may be encountered when, for the first time, a new fixed fund is integrated into a production process. In the first case (top left), despite having the information relating to the working conditions of the unit sufficiently ahead of time, one or more of the components of the fund fail. Such failures tend to be accounted for by the miscalculation of certain operating variables. This may, for example, lead to the rapid deterioration of certain parts or pieces. Another of the cases shown in the above table is when the scheduled operating capacity of the fund does not meet the expectations of the operator. This then would reveal a serious failure in the design stage and would probably mean the redesign of the whole facility.

Table 3. Problems with new fixed funds

		Type of problem	
		Faulty operation	Insufficient operation
Access to the information	Prior to putting in place	Wrong operating capacity parameters	Serious design fault
	After putting in place	Unforeseen factors	Changes to the product manufactured

With respect to the occurrence of unanticipated problems or the failed operation of the equipment due to a totally unforeseen factor, the cause of these may be changes recently introduced into the products manufactured which are not compatible with the capacities of the fixed fund. The two problems are closely related to the very complexity of the various production processes and the environment they operate in. Indeed, the difficulty of obtaining both comprehensive and stable overview increases the risk of problems occurring for unforeseen contingencies.

On the one hand, table 3 has not taken into account non-critical limitations or faults, given that these may go un-remarked, sometimes even for extended periods of time. On the other, neither does it consider the complications that may derive from the complexity of the equipment, especially if the sub-systems are not lineally related. The impossibility of foreseeing all possible contingencies and the fact that there may be several different fault sequences for one single problem mean that an assessment of the reliability of the system would take up too much time and/or an unacceptable amount of human or material resources. A conclusion that is not completely free from arbitrariness and/or subjectivity.

Returning now to the case of relatively simple equipment, once a problem has been detected, it will have multiple levels of resolution. The lowest level is the readjustment of the functions of the fixed fund in order to guarantee a satisfactory level of operation, without attacking the root of the problem. The highest level would mean the complete redesign of the system to eliminate any need for adjustment. An endless variety of options lie between the two extremes.

In a few words, part of the process of learning to operate a fixed fund requires the discovery of latent problems, even when dealing with machinery operating in a stable environment. Given that, on the one hand, faults tend to be detected in the early days, soon after a new unit enters into service, and on the other, the diligence with which the workers will perform their tasks likewise increases swiftly at the start, just a small amount of accumulated production is sufficient for the learning curve to fall quick in from its origin. This curve will rapidly level off almost completely meaning that its impact on unit costs is disappearing. Figure 34 shows a hypothetical relationship between the factor of autonomous learning through practice, the duration of the elementary process and the output per unit of time.

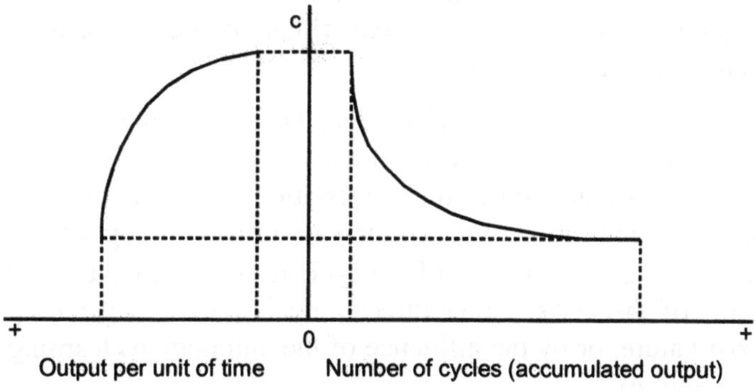

Fig. 34. Experience, cycle and level of production

In the figure, the accumulated amount of output and the working time (or man hours) per output unit has been measured in terms of the production cycle. Then, as can be seen on the right of the figure, the reduction of the duration of the production cycle with a given technique is faster in the early days of production process activity. On the left of the figure, it can be seen that the reduction of the cycle implies a greater amount of output per unit of time.

Despite all that has been said up to now, it must be pointed out that the process of the increasing division of work has always been accompanied by modifications to the fixed funds. Therefore, one cannot separate it from the introduction of a certain degree of technical innovation. The two are responsible for the increasing returns of scale.

The specific features of the process of the division of labour, with respect to technical change, are as follows (Corsi 1991: 4-5 and 42-44, 2005):

1. It is a process subject to the systematic influence of numerous economic factors exerting varying degrees of pressure.
2. Its effects appear discontinuously. This feature accounts for the leaps the process causes with respect to the productivity of work over historical time.
3. It is an unforeseeable and contingent process.
4. It is a path-dependent process: its pathway will depend on the organisational choices made and the productive infrastructure previously installed.

The sum of all these features will determine the dynamic profile of the productivity of work in a plant from a given sector.

Having mentioned some details of the specialisation of work and the implementation of fixed funds, we may return to the main discussion. In terms of the model proposed, an increase in the scale of production may be brought about in one of two ways, each with similar consequences, despite conceptual differences. That is:

1. The acceleration of the pace of the line. That is, cutting the *launch times* of the output units in process without changing the characteristics of the tasks or their assignation to the workstations. In short, no alteration of the internal organisation of the elementary process is triggered.
2. The reduction of the duration of the cycle (c or δ), achieved by changing the times of the tasks, either directly by implementing measures of a Taylorist nature, or by the influence of the autonomous learning process (learning curve).

With respect to the former, see figure 35. It shows a case based on the already familiar elementary process, with 11 tasks performed by 4 workstations (shown at the top of the figure).

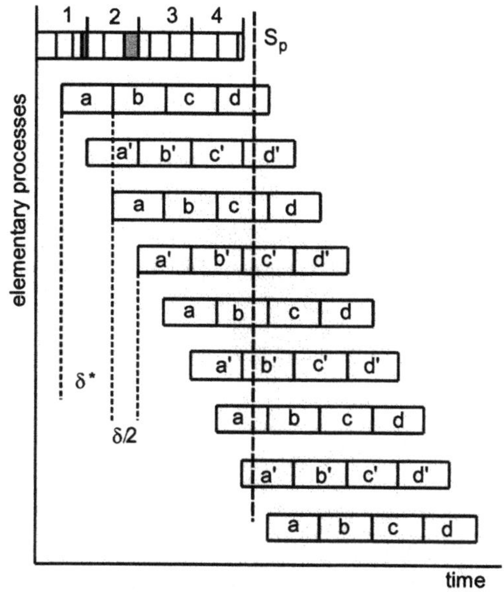

Fig. 35. Increased line speed

With an optimum cycle of δ*=12 minutes, if the pace of activity of the process accelerates to $\delta = \delta^*/2 = 12/2 = 6$ minutes, how many elementary processes will now be simultaneously active in the production unit with the new time lag? The answer is obvious: given that the size of the line process (S_p) in the current productive context is equal to T/δ (or T/c), reducing the cycle by half will double its size. In effect, as can be seen in the figure, the number of simultaneously active processes will increase to eight. Moreover, given that the pace of activity of the funds will not have increased, doubling the speed of the line will then require the number of workstations to be doubled (shown as {*a*, *b*, *c*, *d*} and {*a'*, *b'*, *c'*, *d'*}), and also the number of fund units. The new workstations will either have to be situated between those already existing, or on the other side of the line. Finally, production per hour will rise from 5 to 10 units.

With respect to the above, a clarification must be made: it is assumed that at all workstations the operating time of workers is already fully taken up, i.e., it is impossible to further increase their rate of work. Nonetheless, the model has been built with very rigid assumptions. In fact, in everyday conditions, working times do not tend to be so strictly set. The duration of tasks will depend on the skill and motivation of the workers as well as the control exercised over them. Therefore, an increase in the pace of production is possible keeping the number of workers and fund units. By exercising direct pressure on the operators, without making any change to the working methods, performance could be improved. It should be noted that this strategy conflicts with relatively tight limits, both in terms of intensity and duration, or in other words, conflict with the *endurance* of the operator.

In general, any organisational and/or technical change that will permit the reduction of the time lag between the output units in production will require an increase in the size of the line, the total number of workstations in the plant and the level of production per unit of time. In effect, an increase in pace equal to δ^*/n, $n>1$,

1. Increases the scale of the process to

$$S_p = \frac{\bar{T}}{c} = \frac{c^* \cdot m^*}{c^*/n} = n \cdot m^*,$$

   a figure which will coincide with the number of workstations.
2. Increases the production output to n/δ^* units, assuming that the unit of time used is hours (Tani 1986: 219).

Given that each workstation will only have executed part of its tasks when a new unit from the following elementary process appears ready for the corresponding processing, the number of funds (ϑ) used by the line will also have to increase in accordance with the rise of the pace of the production.

All of this will only be possible if there is no restriction to the supply of flows and funds, no break in the stream of components to be assembled, or limitations of demand to absorb the additional output.

One point worthy of attention is the difference that emerges between the level of production and the time efficiency (E) of the process. Indeed, the line will have increased its level of activity (production per unit of time) without reducing the delays contained into the elementary process. In the theoretical case considered here, the scale grows without any change to the performance of the funds. This is due to the greater number of funds and amount of flows used.

With respect to the reduction of the cycle (c), as a factor in the increase in scale, the analysis is based on the composition of the production activity engaged in by the different workstations, that is,

$$c = s_i + e_i = p_i + \pi_i + e_i,$$

There are two main methods by which this cycle may be shortened:

1. Reducing the effective work (p_i) through, for example, simplifying the production operations of the tasks. In other cases, it may involve changes to the material, such as the standardisation of the parts or components to be assembled, or introducing improvements to tooling or machinery.
2. Reducing the output in process transfer time between the workstations and other factors included in the non-processing time (π_i).

None of the above changes will necessarily involve a reduction to t_{sum} (or the sum of the service times). In effect, a fall in the service times of one or more of the tasks would not conflict with the fact that certain tasks may be of a greater duration, although it is a situation which is none too plausible. With regard to the number of workstations, it is clear that this may *not* be reduced. If c represents the activity cycle, a reduction in this interval would actually mean maintaining or increasing the number of workstations. It would depend on the results of the re-optimising of the line in which the order of tasks would, in accordance with the assumptions established, remain the same. Thus, all that is fundamental is that the resulting cycle is shorter. Not even the remaining idle times (l_i) would necessarily have to be reduced, although of course it is logical to imagine that they would be, given that the elimination of this is a priority objective. On the other hand, it must be borne in mind that a reduction in the cycle automatically means an increase in process speed. Shrinking the cycle is equivalent to a smaller interval of time between consecutive elementary processes.

Among all the features of the new line, the only thing of interest is the (reduced) value of the cycle. The other variables, that is, t_{sum}, T, m, E, and BR are not of importance. Therefore, the greater scale derived from a shorter cycle, given the lag between consecutive elementary processes, would not necessarily be associated with a shorter total operating time, or greater time efficiency.

With the reasonable assumption that a reduction in the cycle is accompanied by a lower c_{min}, the segmented curve that represents the (c, m) combinations would move towards the origin of the coordinates. This movement would be accentuated if m_{max} were reduced by the effects of the new c_{min} and/or t_{sum} values. Hence, an increase in the scale of production, deriving from technical and organisational innovation, gives rise to the (c, m) combinations moving gradually towards the origin of the coordinates.

Figure 36 shows the relationship between the output per unit of time (vertical axis) and the lag or cycle values (horizontal axis).

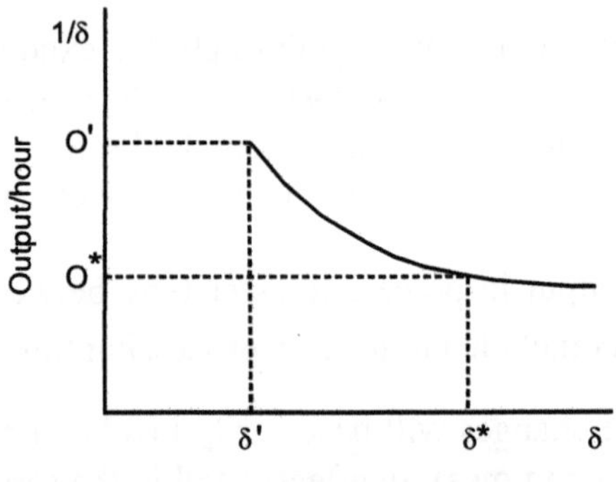

Fig. 36. The relationship between the lag/cycle and the level of production

On the one hand, the general hyperbolic shape of the production per hour curve is accounted for by the fact that it is equal to $1/\delta$. On the other, the δ' cut-off of the curve represents the fact that, without major technical change, reductions to the lag and/or cycle have physical limits. So δ' may be considered as the *minimum* value achievable. But it is a maximum pace of work that may only be tolerable for short periods of time. Thus, $\delta' \leq \delta \leq \delta^* = c^*$. Obviously, technical innovation that alters production elements and procedures will represent the displacement of the $1/\delta$ curve, although its general shape will remain the same. Meanwhile, with a given technique, any increase in production per unit of time will require an increase in the rate of activity, a strategy with clear limitations.[14]

In short, there are two ways of increasing the production per unit of time of a line:

1. The first is the direct reduction of the lag between consecutive elementary processes. That is, $\delta = \delta^*/n$, $n > 1$.
2. The second is to reduce the duration of the cycle contained in the elementary process. This is denoted as $c = \lambda \cdot c^*$, $0 < \lambda < 1$.

Thus the combined effect of the two would give rise to a level of production per hour equal to,

---

[14] In addition, a δ value higher than the minimum achievable offers an undisputable advantage: it permits the operators to perform minor rectifications and make good small incidents in the operating of the fixed funds without having to stop the line for petty problems.

$$\frac{n}{\lambda}\cdot\frac{1}{\delta^*},\ \text{or},\ \frac{n}{\lambda}\cdot\frac{1}{c^*},\ n>1 \text{ and } 0<\lambda<1.$$

Although the two methods are equivalent, they have different effects on the number of workstations and fund units required, and on the layout of the assembly line.

3.3.2 Work teams and the pace of production

As indicated above, since the 1970s, the Fordist line has undergone significant modification. Outstanding is the disappearance of the organisational principle of *one post, one operator, one task*, which is replaced by work teams who share the different tasks of a given stage of the production process. Prominent amongst the advantages that work teams represent for the company is the way they increase the pace of production because of the reduction in idle times. Indeed, workers and production operations may be grouped and reorganised in such a way that they are performed with no loss of continuity. This practice will even permit a reduction in the total number of workers required.

Let us take a phase of a given production process with n workers who perform their activities at individual workstations, deployed in a rigid sequential order. On the one hand, with a set c cycle, the total duration of the phase is equal to $n \cdot c$, that is, the cycle multiplied by the number of workers, or number of tasks. On the other, each workstation has its processing (p) and non-processing (π) intervals and lapses of inactivity (l). Thus, it could be said that,

$$n \cdot c = (n-1)c + c = \sum_1^n p_i + \sum_1^n \pi_i + L' + L^c$$

That is, the total working time available to perform the different tasks of the phase in question is equal to the sum of the three intervals of time considered. It should be noted that the sum of the idle times has been divided into two: L^c is an interval equal to c, while L' indicates the remaining time of inactivity contained in the phase.

It is clear then that,

$$(n-1)c = \sum_1^n p_i + \sum_1^n \pi_i + L'$$

meaning that the grouping together of the different workstations will permit the services of one of the workers to be dispensed with. A situation that will occur, to a greater or lesser extent, as long as

$$\sum_{1}^{n} l_i \geq c$$

This is very simply accounted for: the establishing of a single work team permits some of the possible idle time for balancing delays between the different workstations to be eliminated. This increases yield as the tasks will be performed one after the other without any balancing requirements (even though, in the end, idle time may still be present), and may even permit a reduction in the number of workers required.

3.3.3 Synchronising the timing of the production process

In the real world, the time synchronisation of the line process is never as strict and persistent as is the case in the hypothetical situation proposed. For example, let figure 37 represent a process with its accumulated curves of finished and started output units.

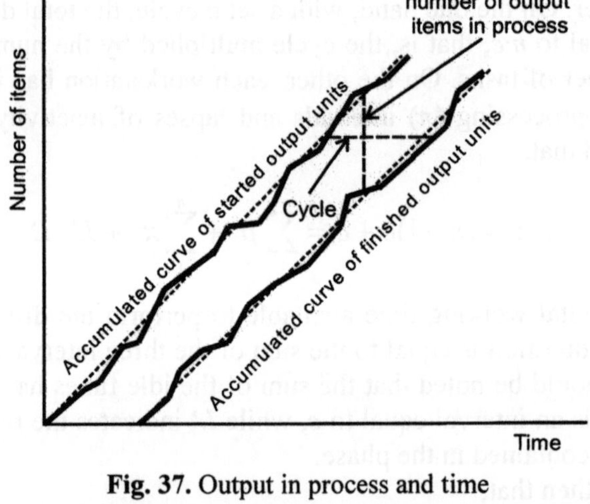

Fig. 37. Output in process and time

The two continuous curves show the number of output units produced over the passage of time measured, for instance, in terms of the number of days worked. As is clear, the horizontal distance between the curves is equivalent to the amount of time needed to produce one output unit, while a vertical reading gives the total number of units in production. Ideally, the two accumulated curves should be straight and parallel (see the dotted lines). In practice, however, it may be assumed that the two curves will twist to a certain extent, at times coming closer together than at others. The output flow curves will be deformed due to the problems of synchronisation and coordination manifest in any real process.

On the other hand, figure 38 shows the effect of organisational change and technical improvement on a given process, *A*.

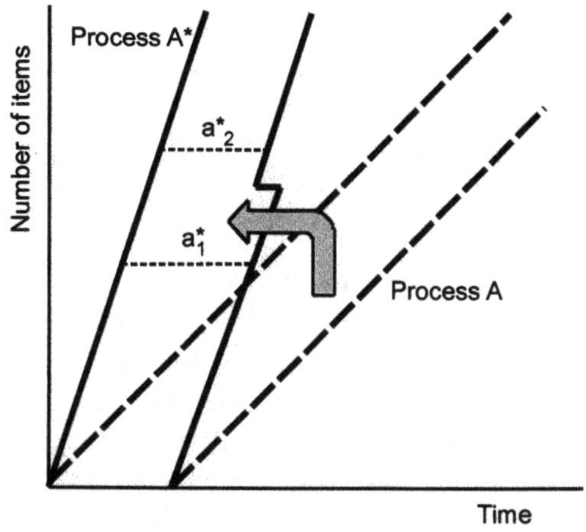

Fig. 38. Acceleration and reduction of process time

Because of technical innovation, the accumulated curves will move towards the *x*-axis (process *A**), thus verifying an increase in the rate of work of the production line. More output units will be produced for each interval of time. The figure also shows the shortening of the elementary process ($a^*_1 > a^*_2$). Both factors will increase the level of plant production. Given the theories around which the diagram has been constructed, any (anti) clockwise movement will represent a lesser (greater) level of time efficiency for the underlying production process.

3.3.4 Changing the production process

As explained above, any drastic elimination of fund idle time will require the elementary process to be deployed as a line process. However, after having met this general condition, the level of production may be increased still further by establishing progressively shorter lag (or cycle) times. Historically speaking, the conveyor belt has been, and still is, one of the ways of increasing production per unit of time: it permits the swifter conveyance of the output in process from one workstation to another and, moreover, obliges the worker to work at the pace it sets. The rate of production will depend on the speed of the belt itself, and on the distance between the units of output it carries. Besides, the so-called scientific organisation of work has progressively shortened the time of execution of the different tasks. Fordism and Taylorism, and their multiple derivatives have framed the historical search for the increase of the scale of line processes. Meanwhile, sooner or later, an increasing scale of production will have to be accompanied by significant changes to the type of flows and capacity of the funds used in the process. In effect, it is sufficient to look at the productivity of work, a clear indicator of the *performance* of any production activity, to perceive a mix of two different factors: first, the rate at which the workers operate or perform and, secondly, the intrinsic technical capacity of the equipment used. The former may be increased without any need to alter the method of production, as stated above. The latter will require the assistance of technological innovation.

The normal activity of a plant may lag behind other plants due to the poorer quality of the material it consumes, limitations to the rate at which its fixed funds operate, excessive idle time between the different tasks or phases of the process, and so on. These and many others are the direct factors which drive companies to seek new production techniques or new designs for the products they make. In general, the results achieved only partially stand out from the procedures and/or elements that already exist. This could occur when firms are not accustomed to the new mix of factors which, technologically speaking, is very different from what they are currently using (Rosenberg 1994: 5). Needless to say, the relative weight of fixed funds in most industrial production methods offers many opportunities for minor improvements to be made. Such gradual change then accounts for the configuration of the technological pathways (Dosi 1982).

Before starting the argument the next assumption must be enounced: the set of new production methods are available to all firms independently of their size.

The funds and flows model is characterised by the exhaustive cataloguing of the different elements that participate in production, along with the complementarity between them. In addition, the temporal dimension of the production processes is taken into consideration. It is for this reason that a high degree of detail could be achieved in the description of what has come to be known as *process innovation.* Indeed, when comparing processes with exactly the same output and identical arrangement of tasks and phases, from a merely descriptive point of view, the replacement of productive elements may take three forms:

1. Change of flows.
2. Change of funds.
3. Substitution between funds and flows.

Obviously there are induced effects fed by the complementarity of the elements, which open up this classification still further. The direct and indirect implications of this tend to make up a dense tangle. And all of this gives rise to the enormous and heterogeneous ensemble of technical modifications to production activity.

In keeping to a descriptive point of view, many of the flows in numerous production processes may be replaced by others. An example would be the case of changing from electricity to natural gas, or replacing an additive with another. Such switches tend to require changes being made to the methods and tooling used. They may also give rise to new varieties of outflow. Although many of these substitutions may be considered minor, there are indisputable exceptions. For instance, the same chemical product may be obtained using very different reagents, but to do so would mean making significant changes to the production procedures.

In the case of funds, their replacement also covers a broad range of options. So, for example, to plough a field, animal power (a team of oxen or mules) may be replaced by mechanical power (a steam engine or tractor). As above, such changes of fund likewise tend to affect the type and rate of consumption of one or more flows (diesel fuel instead of forage), as well as the mode and time of use of funds (a multi-blade share plough to replace the old single blade balance plough). When dealing with the replacement of the services of work by more hours of fixed fund activity, i.e., a greater level of automation of the process, it must be remembered that it will tend to mean an increase in the consumption of certain flows as energy.

The direct replacement between funds and flows is perhaps a less common situation. However, it is possible to find some examples. One of them is the use of dynamite instead of labour for levelling work or to open up a tunnel.

Adding the time dimension, the typology of technical innovation forms expands. Now, it must take into account,

1. The layout of the tasks (or phases).
2. The rate of flow input and/or the service times of the funds per output unit.
3. Instants at which inflows are incorporated.

In addition to the abovementioned innovations, there are others which, on a practical level, may also be of the greatest importance. These are all related to the issues of storage, a function which in many processes occupies a great amount of time and thus has a notable influence on both efficiency and costs. Here then the type of technical change considered would involve:

1. A reduction of the time that goods in process are held as technical stock. For example, tanks which are appropriately refrigerated will help accelerate the ageing process of wine.
2. Holding smaller stocks of input or output flows. This may be achieved with more advanced stock control systems which combine the criteria of simpler spatial location, more potent and/or faster systems of transport and quicker information management systems.
3. On some occasions, innovation may consist of lengthening the period in which goods are stored. Such is the case with cold stores for fruit, the freezing of meat and fish, and so on.

Of course, in certain production processes, one may find a combination of all the abovementioned types of technical change. To all events, it is assumed that the consequence of any such technical change is a relative reduction of unit cost.

In general terms, to have a complete picture of technical replacement, sufficiently clearly identified elements of production have to be dealt with, and never loosing sight of the induced effects of any change. Consequently, to merely speak of capital/labour replacement would be too general and, therefore, misleading. On the one hand, it does not consider the specific traits of the replaced units of those funds. On the other, it ignores the consequences on flow rates: the use of a paint robot will reduce the number of workers needed, as well as the quantity of paint consumed per unit of output, but at the same time will increase the amount of electricity consumed by the process. It is only from a detailed picture of any such technical change that we will be able to develop a truly significant economic assessment of this.

As is known, the implementation of process innovation is faced with significant dilemmas:

1. The wish to benefit as fully as possible from all of the latest technical novelties can breed distrust if these innovations are insufficiently tested.
2. The existence of a high degree of complementarity between the different elements of production, which adds limitations and requires adjustment.
3. The need to fully amortise the existing equipment, although it is obsolete.
4. The loss of the benefits of accumulated know-how and experience with old methods and techniques.

The alteration of the number and/or order of the tasks, and changes to the nature or amounts of the funds and flows employed, originate problems of coordination with the other stages of the production process. These and other obstacles account for the fact that changes to production processes tend to be partial.

Any discussion of the diverse ways in which technical change may occur in the production process cannot conclude without reference to the way product innovation may also indirectly impact on this. The ever increasing improvement to and variety of features of, the finished product will also affect both the number and the complexity of the manufacturing operations required. In effect, adding new features is translated into more (sub) systems and components (such as ABS, catalytic converters, air-conditioning or airbags, in the case of the automobile industry) to be assembled. This addition to the complexity of the production process increases the number of tasks to be performed and, also, the total duration of the process. To cope with such innovations, new funds and flows will have to be incorporated, often requiring a particular effort to be made in terms of design and standardisation, and generating the need to reorganise all the existing tasks to accommodate the new ones.

Taking the conceptual framework developed thus far, an initial attempt may be made to shape process innovation related to the funds and flows. It consists of a description built by comparing the techniques employed by processes of the same kind, identified by an output considered sufficiently homogeneous.

As hinted at above, innovations that have an incidence on the funds may affect their presence in the process in three different ways, due to the twofold way in which their services are measured. Indeed, the activity of any fund, be it with respect to a task, phase or the process as a whole, is measured by the number of units employed multiplied by the time of operation. Therefore, the technical modification of a fund, regardless of its specific nature, may represent a shorter interval of activity, or maybe a reduction in the number of units required, or both. Thus, the evolution of the number of fund hours for the different techniques could be as shown in figure 39.

The $\alpha\beta\gamma$ pathway indicates the changes in the number of units of a hypothetical fund (as long as it is acceptable to consider the use of the same fund) and in its times of activity, both measured per output unit (supposedly homogeneous) due to technical innovation. Such a diagram could represent, for instance, one of the chief consequences of technical change: the reduction in the amount of direct human labour contained in manufactured products.

With regard to flows, successive technical changes could lead to a reduction of the input rate per unit of output. This saving in the amount of these inflows (energy, water, raw materials, etc.) may be quantified by collecting data from specific process innovations. The time pathway of this reduction can be highly diverse. As a hypothetical example, figure 40 shows the reduction in the amount of a certain type of flow used per output unit over time (stepped line).

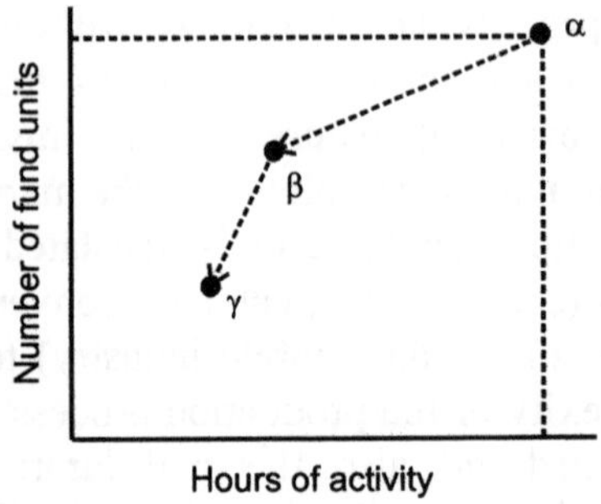

Fig. 39. Fund hours and technical change

In this case, if f_k is the technical coefficient of the k (k=1, 2,..., K) input flows that are all of the same type, the time pathway (successive techniques) of the unit consumption of these may be estimated with the following function (continuously decreasing curve):

$$f_k(t)=f_k(0)\, e^{-\alpha t},\ \alpha>0.$$

The function describes the progressively decreasing amount of a given flow used per output unit by a single family of production processes over long periods of time. Evidently, other forms of decreasing pathway would be equally plausible: lineal, a progressively accentuated decrease and so on.

Any representation of changes in fund services and flow rates should not only relate the direct, but also the indirect effects. A complete picture of both effects would be highly complex to draw. Note that *anomalous* cases could easily be found: those in which process innovation has led to an increasing need for one or more elements and thus will present pathways that grow over time. A good candidate for such a paradox is, for example, the direct consumption of electricity in the production of many goods: a flow rate that increased steadily over a large part of the 20th century.

Finally, figure 41 provides a theoretical example of the changing levels of production per unit of time brought about by process innovation. As can be seen, the appearance of new technique β represents, in comparison with the current one α, a forward leap in terms of the technical productivity of the fixed funds. This new technique β offers a normal rate of activity, δ^*_β, superior to that of the old technique, δ^*_α: $O^*_\beta > O^*_\alpha$. However, the maximum performance level with method α is higher than the normal level offered by technique β: $O'_\alpha > O^*_\beta$. This fact may act as a brake on the introduction of the new technique.

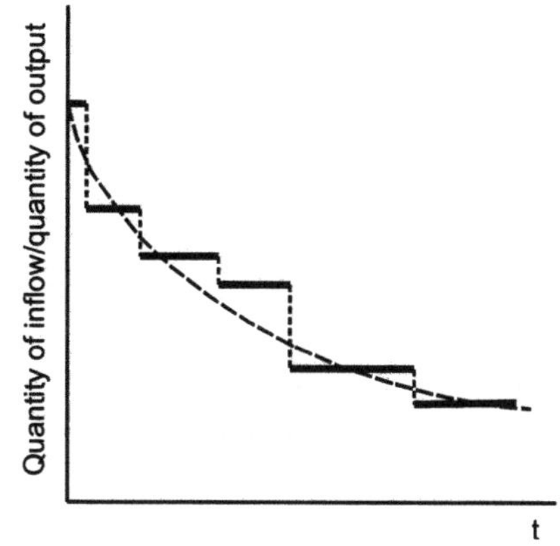

Fig. 40. Amount of flow *k* per output unit

What is of most interest in figure 41 is that it allows the intervals of variation of the performance of the funds for each of the two methods of production to be distinguished. This leap in productivity may be measured by the number of output units in production that may be processed per unit of time by the new technique in comparison with the old one.

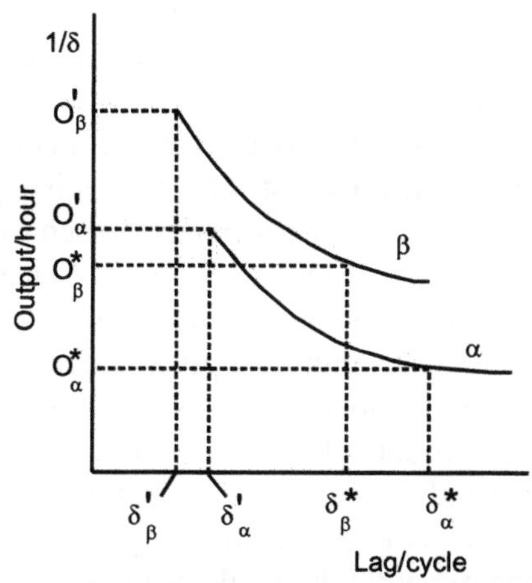

Fig. 41. Level of production, lag/cycle and technical change

3.4 Indivisible funds

Generically, indivisibility is the attribute possessed by everything that looses functionality if fragmented. This means that if something is divided it will be rendered useless. Besides, if it is wished to replicate it, it must be done by multiples of integer numbers. Indivisibility is a physical affair that could be observable at very different degrees and with some significant economic implications.

With respect to the *indivisibility of the process*, there are two kinds of boundaries which should be taken into account:

1. The threshold that has to be reached to start it up.
2. The minimum threshold required to keep the process operating, without changing its nature.

The two levels of production do not need to be the same. As is expected, technical innovation may affect both the lower and, if applicable, upper limits.

In general, the minimum level of activation required for a production process is the output unit. This will often be a level that is technically feasible but economically non-viable. It is the case, for example, of many chemical processes: absolute minimum amounts for inflows (reagents) may be required to achieve a sufficiently high critical mass to trigger the reaction envisaged (given a certain temperature, pressure, etc.) provided that these inflows are maintained in strict proportionality. In such cases, although the substances that participate in the reaction are divisible, the process as a whole is worth of defining as indivisible since the combinations of such inflows involve set amounts. It may also be the case that once the process is started, it can be sustained with lower levels of flow than initially required for activation.

There are processes that have one or more interim stages (tasks or phases) the economic dimension (cost) of which is independent to the scale at which the rest of process is later executed. In such cases, indivisibility is defined more in economic than technical terms. For example, this would be the case of the production of the first copy of a book, or the construction of a prototype. Such tasks require a given amount of both time and resources, regardless of the number of copies that will later be published, or the number of finished products finally produced. Even though the cost of this phase may be reduced by technical innovation, the same would generally occur to the process as a whole, meaning that its relative impact, inflexible from an economic outlook, would remain more or less the same.

With respect to *funds*, the property of *indivisibility* is found extensively. In reality, when indivisibility is mentioned within the context of production analysis, the interlocutor will almost always be referring to funds (that possess that quality). Fund indivisibility may be classified in one of two ways (Morroni 1992: 26):

1. A fund element will not be usable for the purpose for which it was designed unless it is complete. Or, if it is divided into parts, each of them is useless taken individually. Examples of this appear somewhat banal: the component parts of a locomotive, a car or an aircraft are useless as a means of traction. They will become a fund only as a whole machine that comes off the assembly line. It should be pointed out that we are referring to the fractioning of a complete unit into its pieces or constituent components, whatever its physical dimensions.
2. All funds have their own specific production capacity, the partial use of which points towards inefficiency. There are many evident examples of this: any means of transport only half laden, underused multi-purpose equipment and so on.

It must not be forgotten that all tooling, albeit the most simple, is a system. On occasion, it is designed to intervene in several different elementary processes at the same time. Therefore, the indivisibility of funds may be defined on two juxtaposed planes: they have to maintain their physical integrity and, at the same time, fully exhaust their productive capacity.[15]

Funds of capital equipment are characterised by their operating aptitudes, measured in terms of an appropriate variable. Their working capacity is determined by technical factors, subject to variation for successive innovations. Then the existence of a relationship between performance and cost could be imagined, delimited by lower and upper limits. The former would include those levels of activity which are technically feasible but economically ruinous. In such a case, the lower is the level of operation the greater is the room for idleness. The upper limits refer to levels of operation which go beyond the technical possibilities existing at a given moment in time. There is no doubt that such upper limits are more clearly defined than the lower ones. Moreover, technical innovation pursues a twofold goal: to extend the fixed fund's capacity by raising the upper limit, while

[15] As is evident, the concept of indivisibility referred to herein has nothing to do with the size or amount of the finished consumer goods sold, i.e., under normal circumstances commercial practice establishes that eggs are not sold individually. This is a form of indivisibility which does not fall within the sphere of production.

at the same time there are devices which are designed to be operative at low levels of activity thus avoiding the presence of excess capacity.

Nevertheless, from the point of view of the analysis considered in these pages, the indivisibility of the funds is a property which is expressed as *multiple capacity*. Thus, a perfectly divisible fund would be one with a working capacity capable of constantly adjusting to the workload of the moment. A quality which could always be stated of the inflows, given that they will always adjust to the requirements of the process unless, of course, the supply is limited or there are technical problems with respect to their incorporation. Obviously, being measured by a quantity of time, services of any fund are always divisible. For example, both the services of the land, that is, a space suited to planting or placing other objects and the working time are divisible. But, if indivisibility is understood as a multiple capacity of funds it could happen that the level of operation achieved in a given productive process does not set its productive capacity. An excess of capacity will result and, in turn, it may generate direct and opportunity costs.

The question of the degree of use also involves technical change. Indeed, certain innovations have enabled the capacity of funds to adjust to the level of operation of the firm, even the smallest. In this way great economic losses derived from excess capacity are avoided. This is the case, for example, of agricultural traction machinery. The development of the internal combustion engine at the end of the 19th century permitted the construction of far more versatile and more powerful smaller tractors at prices far more affordable to most farms. This left behind the enormous, expensive, and clumsy steam engines that had been employed up to that time. At the same time, farm implements were also adapted to meet the requirements of different sized farms. However, in those regions with small family farms, it is possible to see the phenomenon of the twofold over-mechanisation already mentioned in the previous chapter. In effect, on the one hand there are many implements and machinery that are only used for a few hours each season. There is an excess of capacity although, to a certain extent, the excuses of simultaneous needs by farmers, along with a very small time slot in which to carry out many of the harvesting operations, are valid. On the other hand, due to the diversification and rotation of crops, a large number of machines are required.

As has already been mentioned, a given fund has multiple capacity if it is able to operate simultaneously on two or more output units. Such multi-task capacity is a quality possessed by many funds. For example, there are machines capable of processing, painting, cutting, cooking, and so on at the same time on several output units. If such funds act on a lower number of elementary processes than their full potential, it will lead to an excess of capacity and to inefficiency and extra costs. The following paragraphs are devoted to the integration of the indivisibility quality of many fixed funds to the fund-flow model.[16]

To help understand the problem of adjusting the capacity of fixed funds, see figure 42.

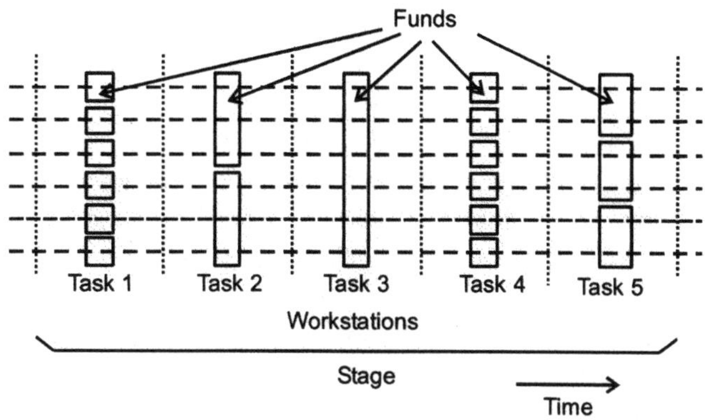

Fig. 42. The line process and indivisible funds

In this case, a particular phase comprises five tasks and workstations, with several funds of specific capacities, represented by the differently sized rectangles. To take up the capacity of the funds fully, six elementary processes have to be performed at the same time. Each of the processes, or output units in production, is represented by a horizontal line. This gives us a parallel line process.

Multiple capacity may be tackled in the following way: let it be said that $\kappa^*_j \geq 1$ ($j = 1, 2, ..., J$) is the maximum number of elementary processes that a j-type fund may process simultaneously. Then the index of capacity used (γ_j) is defined as the ratio (Petrocchi and Zedde 1990: 67),

$$\gamma_j = \kappa_j / \kappa^*_j \leq 1$$

κ_j being the number of output units on which a specific fund is working in a given process. As γ_j approaches one, excess capacity falls. Thus, the cited index indicates the degree to which the capacity of a fund is saturated.

In order to guarantee that a parallel line process operates without any fund unit experiencing excess capacity, the number of units of a j-type fund should be equal to the *minimum common multiple* (henceforth MCM)

[16] Nonetheless a process may be envisaged in which an operator, at the same time, views several units of a (semi) finished product, as an initial quality control procedure. With respect to indivisibility, see Landesmann (1986), Morroni (1992: 25-6), and Piacentini (1996).

of its κ^*_j divided by the intrinsic capacity of that kind of funds (Petrocchi and Zedde 1990: 70). That is,

$$\tau_j = MCM/\kappa^*_j$$

In the case shown, $\kappa^*_1=1$, $\kappa^*_2=2$, $\kappa^*_3=3$ and $\kappa^*_4=6$. Thus, the MCM is 6. This means that 6 units of type one fund, 3 of type 2, 2 of type 3 and 1 of the last type will be required for the process to operate continuously and properly.[17] The question is one of aligning six elementary processes in parallel.[18]

The term τ_j leads to the concept of the *multiple elementary process*, i.e., to overcome the inefficiency created by the various capacities of the funds, it is necessary to deploy several elementary process in parallel in a number equal to those needed to eliminate the idle time of the fund with the greatest capacity.[19] Therefore, the scale of a line process depends on two variables: the size of the hypothetical multiple elementary processes executed in line and the time lag between them.[20] See figure 43.

It is the interaction of the above two factors that accounts for the large scale of production. This may be due to the span of the multiple elementary processes, along with the greater number of single elementary processes completed at the end of the working day. Obviously, the scale of production will undergo a huge expansion if its span is drastically increased and/or the time lag between consecutive multiple elementary processes is shortened. Therefore, every production technique may be associated with a given combination of the degree of capacity of the funds and the intensification of the process per unit of time. Any innovation will then involve a variation in both variables. This change of scale will probably be

---

17 For the sake of simplicity, it is assumed that each task only uses funds of the same kind. However, once the issue of the coherence of capacity and the number of elementary processes is resolved, this assumption may be dispensed with without detriment to the integrity of the model.

18 If all the fixed funds were simultaneously active in all the elementary processes, then $\gamma_j = \tau_j = 1$ as $\kappa_j = \kappa^*_j = N$ (number of parallel processes).

19 Given τ_j, the degree of adjustment is the ratio $\rho_j = \tau_j / \tau^*_j \leq 1$ (Petrocchi and Zedde 1990: 72). Each ρ_j measures the fine-tuning amongst the number of units in each of the different kinds of funds used in the elementary process, and the number of these contained in the multiple elementary process. As can be seen, in a balanced multiple elementary process $\rho_1 = \rho_2 = \ldots = \rho_j = \ldots = \rho_J = 1$.

20 The example assumes the total simultaneousness of the multiple elementary processes. That is, each and every one of the successive phases coincides in time. Nonetheless, this is a supposition that may be relaxed.

accompanied by the use of new materials (flows), changes to the content of the tasks and so on.

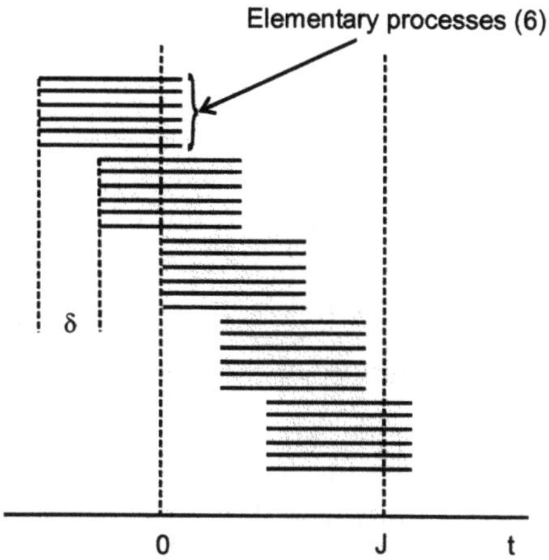

Fig. 43. Multiple elementary processes

Having reached this point, the above reflections on the issue of scale may allow some of the qualitative features that distinguish sequential from line process to be specified. Namely (Scazzieri 1993: 95):

1. The scale of the activity carried out in a craft workshop is most closely associated with the degree of skill of the craftsmen. The greater the level of skill, the greater the number of units of the same elementary process that will be executed during the course of the working day and, likewise, the larger the amount of different types of elementary processes that could be performed. The main way of extending the scale of operation will be via skills training.
2. In a line process, the scale of the operation will depend on the performance of labour and the degree of automation. The two factors allow the shortening of the lag between an increasing set of elementary processes simultaneously deployed in line. Each scale will employ its own particular technique, chosen from amongst the range of production methods available for the process in question. Going further still, it may be said that the goal of altering the scale of a process does, in itself, favour the emergence of new production and/or management techniques. This will be an attempt to generate what certain authors refer to as *dynamic economies of scale* (Morroni 1994; Scazzieri 1993). Thus, changes of scale imply the alteration of the technical coefficients of the productive process.

The existence of multi-capacity funds varies with the size of the line process (S_p). Now the number of simultaneously performed elementary processes will be affected by the aforementioned minimum common multiple. That is,

$$S_p = T/\delta^* \cdot MCM = m \cdot MCM$$

m being the number of workstations. As may be assumed, for each different kind of fund the number of physical units in operation (k_i) will be τ_j. Finally, with regard to the level of production, the combination of all the aforementioned factors will result in:

$$\text{Output per unit of time} = MCM \cdot 1/\delta^*$$

In effect, the rate of production would correspond to the rate of a single elementary process multiplied by the optimum size of the multiple processes (MCM). Figure 44 illustrates the values of δ along the horizontal axis, while the vertical axis indicates the output per unit of time. Ceteris paribus, the optimum activation of a production process with multiple capacity funds will lead to a higher level of production. Therefore, a higher curve appears which is convex with respect to the origin of the coordinates and parallel to that of the deployment of an individual process. In the figure, the higher level of production (*O''*>*O'*) is proportional to the adjustment value for the funds of different capacities.

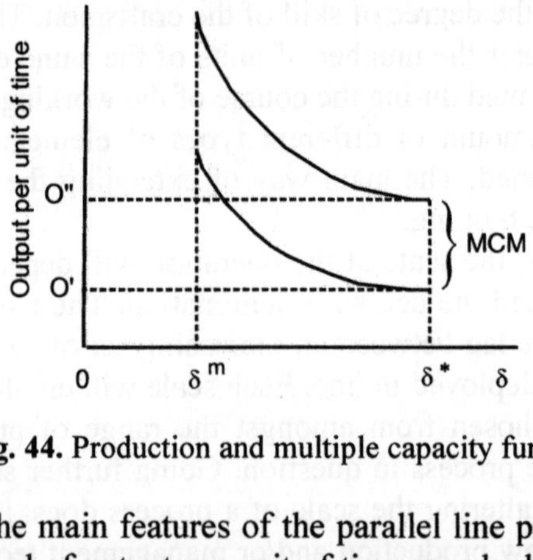

Fig. 44. Production and multiple capacity funds

After explaining the main features of the parallel line process, all the factors of a higher level of output per unit of time may be brought together in a single expression. That is,

$$\text{Output per unit of time} = \frac{n}{\lambda} \cdot J \cdot D \cdot MCM \cdot \Lambda \cdot \frac{1}{\delta^*},\ n>1,\ 0<\lambda<1 \text{ and } MCM>1$$

in which $1/\delta^*$ may be considered as the theoretical basic level of production. Therefore:

1. The factor (n/λ) shows the reduction of the cycle brought about by the use of Taylorist and Fordist methods.
2. The two following terms indicate the duration of the working day and its number per annum ($H=J{\cdot}D$). Or, put in other words, the total working time per annum.
3. The MCM value represents the coordination required for multiple capacity funds.
4. Scalar Λ indicates the number of times the multiple elementary processes is replicated in the production unit.

In the case of the latter, it must be remembered that a production line may be replicated to increase the level of production of the plant. Given that a multiple elementary process is an efficient form of deployment, any multiple will also be so (see figure 45).

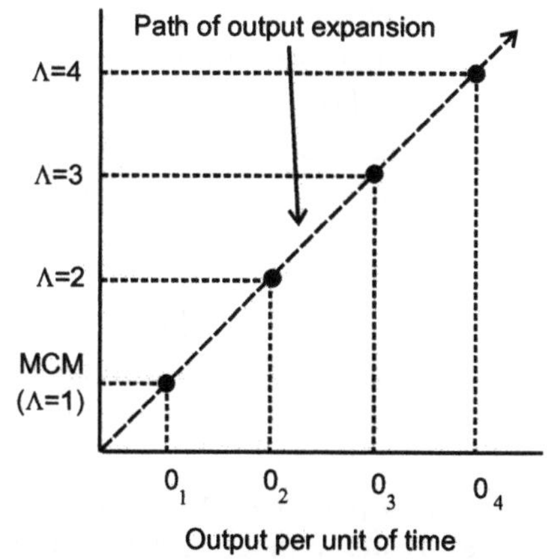

Fig. 45. Pathway of the increase of scale

Once again, the figure illustrates Babbage's multiple principle that establishes that, to increase the scale of a balanced process in line, this must be replicated an exact number of times. Babbage's principle can be included in the model by way of parameter Λ.

The level of production will increase with parameter Λ do. In this respect, figure 45 indicates the significant points that make up the line along which output expands. These points indicate the replication of a hypothetical multiple elementary process. The increasing levels of output per unit of time (O_1, O_2, ...) have the same level of average costs. Between two consecutive points, the cost per unit will be higher due to the fact that there will be funds operating with excess capacity.

Summarizing, the existence of one or more indivisible funds in a given production process in which, as is known, there is a predominance of complementarity relations between the inputs involved, give rise to its expansion following a path of discrete steps (or multiples of the optimised basic scale). Therefore, the need to combine the production capacities of the different machines emerges as a key factor in accounting for the size of a production process and thus the size of the production unit.

3.5 The production process and its size

In this chapter different concepts referring, one way or another, to the *size* of production activity have been discussed. To avoid confusion, there is a list of all of the concepts below. These are:

1. The *Size of the line process* (S_p) or number of elementary processes simultaneously activated. When a multiple elementary process is active, S_p increases proportionally. The same occurs if several production lines coexist in the same plant. If the output of such lines is highly diverse, ambiguity is avoided by considering these specific facilities as different individual plants.
2. The *Scale of a process* is the amount of output per unit of time. Obviously, the concept refers to one single kind of output, although it may also apply to batches with only slight differences. This is a variable which is directly affected by the indivisibility of the funds, the rate of work, the length of the working day, and so on.
3. The *Capacity of a plant* or *production unit*, resulting from multiplying the scale of the process by the number of hours of activity (per annum).
4. The *Size of a firm* is measured, given the heterogeneity of its products, in terms of value. The firm's non-recurring receipts will often have to be added to this to obtain the total turnover of the company. Logically, any change in the size of the company may be the result of a multitude of factors. Therefore, relating it merely to the efficiency of production activity is quite misleading.

The indiscriminate use of terms relating to the magnitude or dimensions of a production process or plant gives rise to ambiguity and much confusion.

4 The fund-flow model and the production function

4.1 The concept of production function

The concept of the production function has existed for many years in the field of economic analysis. It started out life as a mutatis mutandis extension to the conceptual corpus related to utility maximisation.[1] At the beginning of the 20th century, the production function started to be used in empirical research with the confluence, on the one hand, of the agronomical tradition for the calculation of response curves and, on the other, the beginnings of the theory of economic optimisation. Over time, the production function concept has been used in many theoretical and mathematical developments, always seeking formulae to aid the producer to make rational decisions in the quest to maximise earnings. The massive estimation of production functions was consolidated in the 1950s with the advances in econometrics and calculating systems. Nonetheless, in the following decade, the model was placed in doubt due to the problems of consistency encountered with the theory of marginal productivity. Despite this, the production function concept has always been widely used and disseminated. It is sufficient to open any manual, or glance through the content of the most economic journals to prove this.

Exclusively for the purposes of this chapter, table 4 classifies the areas of research in which the production function has been employed. This arises from the combination of two criteria: the degree of aggregation of the data used and whether the nature of the research was conceptual or applied.

---

[1] The idea of representing technology as a field in the area of merchandise in the same way as consumer preferences, was initially suggested by a certain H. Amstein to Walras in January 1877. Presented as a metaphor, the idea did not prosper. Years later, the idea was to be given due credit with Wicksteed, Barone and, most especially, Johnson. It was the latter who invented the term production function, linking it explicitly to the pure utility theory (Mirowski 1989: 309ff.).

Table 4. Type of production functions

		Scope of the model	
		Macroeconomic	Microeconomic
Object of the research	Theoretical	Analysis of the growth and distribution of income	Criteria of optimum assignation and entrepreneurial rationality
	Applied	Estimation of aggregate production functions: the contribution of the factors and technical change to growth	Assessment of the efficiency of the production process

The definition of a production function, especially in microeconomic research, tends to be accompanied by a long list of assumptions. The most common are the following (Frisch 1965; Ferguson 1969; Johansen 1972):

1. Output is single.
2. The process of transformation is instantaneous: all the inputs intervene at the same initial instant and the complete output emerges immediately.
3. Both the product and the factors may be measured in terms of technical units or, in default, by indices of a quantitative nature.
4. The technique is constant. Consequently the combination of the factors takes place within a given framework of technical knowledge.
5. The production factors specified are continuous and thus there is an infinite number of possible combinations of them.
6. The inputs are totally separable. Or, in other words, there is absolutely no degree of complementarity between them.
7. The functional form is differentiable twice.

The above assumptions is supplemented by the supposition that there is an agent who, given certain preferences, makes decisions to maximize benefits in an environment of perfect competition.

On the basis of the above assumptions, numerous formats for the production function have been proposed. The most generic defines the function in terms of the following expression,

$$Y = F(V_1, V_2, ..., V_n)$$

in which a static framework, $V_1, V_2, ..., V_n$ indicates a certain quantities of factors, Y denotes the amount of product and F the function linking the two magnitudes. This definition will often be juxtaposed with another, establishing the functional relationship with flow rates,

$$y = f(v_1, v_2, ..., v_n).$$

here factors (v_1, v_2, ..., v_n) and the product (y) are amounts per unit of time.[2]

The most widely used mathematical specification of production functions was that proposed by Cobb and Douglas at the end of twenties, although there are indications of an earlier version first established by Johan Gustav Knut Wicksell (1851-1926) in 1895.[3] Nowadays, the Cobb-Douglas version is merely seen as a particular case of the more general CES format.

As is commonly known, Cobb and Douglas (1928) used real data on the US economy between 1899 and 1922, to test the hypothesis of the distribution of income empirically, based on the marginal productivity of factors. Although the econometric adjustments were excellent, the model was soon to be questioned (Mendershausen 1938). Years later, many authors challenged the methodological and theoretical consistency of the Cobb-Douglas aggregate production function, placing its pretensions as a *law* of economics in serious doubt.[4] These objections may equally well be applied to all the variants of this forerunner model.

Representing a production process by way of a production function is a highly simplistic approach: it ignores both the material characteristics of the elements that participate therein and the deployment over time of the activities of the process, as well as the relationship that exists between the process and the environment in which it takes place. Thus the production process is represented as a black box, as shown in figure 46.

Fig. 46. The production process as a black box

A black box approach is commonly found in poorly developed scientific models and theories in which the subject of study is dealt with as if it were a simple unit, ignoring its internal structure. The only interest is in the

---

2 The two concepts have been considered equivalent (Frisch 1965). See Georgescu-Roegen (1969, 1972: 280, 1986, 1988, 1990) in which it is demonstrated that this is not the case.

3 See Sandelin (1976). However, there is an earlier version. This can be seen in the work of Johann Heinrich von Thünen (1783-1850), as argued by Heertje (1977).

4 There are solid studies which reveal that the aggregate production function is a mere dynamic expression of the identity of the national accounting system (Fisher 1971; Shaikh 1974, 1980, 1988; Simon 1979; Lavoie 1992: 36). Thus, this function would not account for the stability of the make up of national income over time, but vice versa, the latter would account for the former.

most directly observable behaviour (Bunge 1998). In itself the use of a black box approach is not rejectable as long as it is remembered that an important goal of scientific endeavour is to reduce the degree of opacity of all models.

In the case of the production function, as stated, it is assumed that the period of production is instantaneous. This assumption prevents the observation of the stages of the process: it brings together all of the inputs in the same initial instant, while the output emerges complete at the end. The immediacy between the extremes implies that the process is wanting in structure. From a factual point of view, this approach brings about the risk of formulating a superfluous model. The well tried cookbook recipe analogy could *not* be applied to this model. Although the production function, like a cookbook recipe, is based on a list of ingredients which need to be mixed together, the simile is imperfect in a basic way: all recipes also contain references as to the procedures to follow to prepare the dish, i.e., they indicate the right times to add the different ingredients, together with the cooking times. Moreover, all recipes give the proportions of the different ingredients. Although there may be some room for manoeuvre, at the cook's discretion, it is clear that if the departure from the established proportions is too great, a different dish will be produced. This is something that is ignored by the theory of production. In effect, the isoquants are constructed by grouping together combinations of closer, but also very distant, quantities of factors, yet all supposedly referring to the same output.[5] Although the physical volume of the output may be the same, a significant change in the make up of the diverse ingredients will give rise to different products, that is, with different markets and prices. This is something any manager is very well aware of. Therefore, a doubt is placed on whether all of the points on a given isoquant refer to the same production process. In reality, a specific production process will only be associated with a portion or stretch of points. Thus the isoquants come to be comprised of small lumps of adjacent points (Atkinson and Stiglitz 1969), or also only by isolated points if the process is characterised by strictly fixed coefficients (Koopmans 1951), as shows figure 47.

The figure shows two different processes which combine the same factors to produce a given amount of two different outputs. Points *a* and *b* correspond to the most efficient methods of production for each of these, given that they require the minimum amounts of one or both inputs, with respect to any other process to be found within regions *A* and *B*. Such inef-

[5] This is not merely an abstract stratagem but appears in isoquants constructed on the basis of real data, as is the case in the example of the process of manufacturing hazelnut chocolate given by Frisch (1965).

ficient processes are represented by points which are close to the ends *a* and *b* within the boundaries delimited by the rectilinear segments. In short, any isoquant map may contain empty spaces due to the limited substitution of the factors in order to maintain the homogeneity of the product.

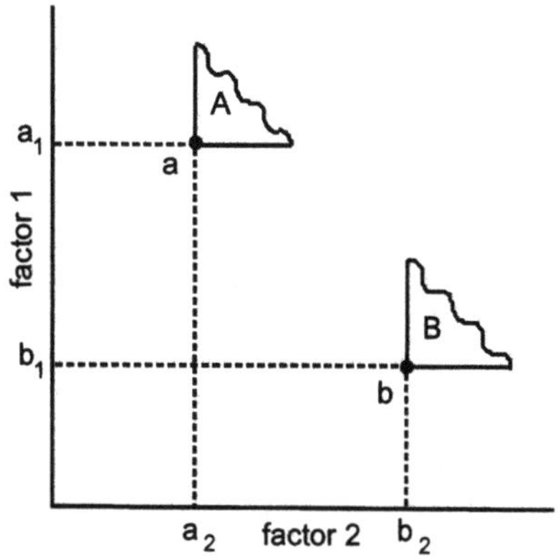

Fig. 47. Isoquants and methods of production

On the other hand, the production function is too conventional. There is a doubtful *ad hoc* proliferation of assumptions employed in its design. Instead of staying within the bounds of the scientific and technological knowledge accumulated in economics and other related disciplines, the model appears to be shielded by a multitude of assumptions which constitute a fictional version of both the nature of the process and the elements involved. Needless to say, such assumptions (continuity, double differentiability, and so on) do have the virtue of guaranteeing results according to the axioms of the rational optimization calculus. All the acceptable mathematical recipes include these suppositions. Indeed, models must show certain returns of scale, patterns of the substitution of factors, and other suitable properties. Thus, making estimates on the basis of real data consists of quantifying the parameters associated to variables, the economic behaviour of which has been prefigured in the mathematical format. In fact, these latter have been chosen for being economically well-behaved.[6] Thus, the specification of the function will depend on the conclusions to be drawn. Conventionalism manifests itself as an artifice and leads to the drawing of priori conclusions.

When dealing with the estimation of microeconomic functions using data from real production processes, the slight differences in the economet-

6 The reiterated practice of deciding the format of the function by convenience has been classified as a *ceremonial* supposition (Hodgson 1993).

ric adjustments are a mere signal of the way different companies use the same technology. The inputs are combined in very similar proportions per output unit, given that each producer uses their own particular simple variation of the same technology. Nevertheless, these results indicate practically nothing about the specific differences in the process characteristics.

Another important issue with regard to the production function is its supposedly instrumental nature since it is based on data supplied by engineers.[7] In this case, empirical data would describe technically feasible production plans and, if represented by a suitable functional form, will enable the researcher to calculate the economic optimisation values. Indeed, these functions would have the following main uses (Varian 1986: 203):

1. The prediction of the level of production as the amount of the inputs changes.
2. The calculation of the marginal productivity of the different factors specified.

Therefore, the production function would derive from the fortunate encounter of response functions as calculated by the engineers, first of all agrarian ones, and the conventional theory of production which, despite having come into existence as a pure abstraction, would have strong instrumental leanings.[8] For example, the functions of how harvests respond to a dose of fertiliser, formally justified by the continuity of the growth of plants and animals and the divisibility of many of the inputs, ended up being transformed into economic production functions, benefiting from the avalanche of analytical concepts and rules of optimisation revealed by economic theory.

It must be pointed out that, from a historical perspective, agrarian engineering was the first empirical frame for production functions reference. To start with, the relationship between agronomy and microeconomic theory, one has to look back to the work of the German agricultural chemist

7 For example, Frisch (1965) considers the work on how crops react to different amounts of fertiliser as a precursor to the economic concept of the production function. Another case is the pioneering work by Chenery (1949) on productions functions related to the pipeline transport of gas.

8 Since the middle of the 19th century, economic theory was progressively concerned with the analysis of maximisation with restrictions. Despite being acceptable for dealing with the very specific issue of the assignation of resources, these new conceptual tools were rapidly applied to all practical economic problems. The result was that the postulate of maximization was elevated, according to its advocates, to the category of *universal law*.

Justus von Liebig.[9] This author enunciated three principles on how crops would react, in terms of production per unit of surface area, to differing amounts of nutrients in the soil (Paris 1992: 1020):

1. The nutritive substances contained in the soil may *not* substitute each other. This postulate is known as the *Law of Minimums*: plant growth is limited by the nutrient of which there is the smallest amount in the soil, even though this latter may contain an excess of the others, in addition to water. Therefore the yield would respond positively to an increased amount of a fertiliser containing this nutrient until another became the limiting factor. This postulate, despite not tackling the issue of the complex interactions among the different nutrients, offers a good macroscopic approach to the relationship between vegetable growth and the chemical richness of the soil.
2. A simple, direct relationship may be established between harvest levels and the amounts of nutrient in the soil or, in other words, between the limiting factor and limited output. Many have interpreted this postulate considering this relationship as lineal. However, on a close reading of von Liebig's work, nothing is found to prevent it from not being so.
3. The response curve reaches a *plateau*, i.e., the yield will no longer respond to increasing doses of fertiliser.

It must be pointed out that his postulates were developed within the context of single crop farming, with the only goal being to establish the maximum amount of output per unit of surface area. Other possible purposes, such as the preservation of the regenerative capacity of the land, are deliberately left aside. It is not a question of disdaining other possible priorities quite simply they are momentarily left out of the horizon of the analysis.

Since the beginning of the 20th century, several different mathematical formulae have been proposed with respect to the principles of fertilising developed by von Liebig. In 1909 the agronomist Mitscherlich presented the following exponential expression:

$$y_i = m\left(1 - ke^{-\beta x_i}\right)$$

in which y_i denotes the production per unit of surface area, x_i the amount of nutrient and m the *plateau* to which the harvest per hectare tends asymp-

[9] In 1840, Justus von Liebig (1803-1873) published his influential work *Die organische Chemie in ihrer Anwendung auf Agrikulturchemie und Physiologie*, in which basic aspects of plant nutrition are detailed thus establishing then cornerstone of agricultural chemistry.

totically. In the above equation, β is a constant that expresses the rate of fall of marginal productivity. As can be seen, the model implicitly admits that the remaining nutrients are not affected by limits.

The mathematician Baule extended the model to deal with two or more nutrients. That is,

$$y_i = m(1-k_1 e^{-\beta_1 x_{1i}})(1-k_2 e^{-\beta_2 x_{2i}}),\ldots,(1-k_n e^{-\beta_n x_{ni}})$$

in which $x_1, x_2, \ldots, x_n$ represent the fertilising substances.

In the mid-nineteen twenties, very simple lineal equations were formulated, such as,

$$y_i = ax_i, \text{ or } y_i = c + ax_i$$

a being a constant of the strict proportionality of the amounts of fertiliser considered and the amount of harvest obtained, and c the corresponding level of production without fertiliser. After the Second World War, such lineal functions became a little more sophisticated, with the proposal of equations of the following kind,

$$y_i = c + \prod_{i=1}^{n} a_i x_i$$

in which y_i is the output per unit of surface area, c the harvest obtained without fertiliser and a_i the coefficients of transformation of the amounts x_i of the fertilisers used together.[10]

More recently, revisiting the von Liebig's work in detail, the following generic formulation of the expression was suggested (Paris, 1992: 1021):

$$y_i = \min\{f_N(N_i, u_{Ni}), f_P(P_i, u_{Pi}), \ldots, f_K(K_i, u_{Ki})\}$$

where, for example, N, P and K are nitrogen, phosphorous and potassium, respectively, and the terms u_{Ni}, u_{Pi} and u_{Ki} refer to the experimental error present. Functions indicated in the equation may or may not be susceptible to lineal specification, and there are diverse assumptions relating to error (additive, multiplicative, or a combination of the two).

---

[10] A great variety of both technical and economic production functions can be found in Heady and Dillon (1961) and Johnson (1970). It should be stressed that Woodworth (1977) gives a long list of 185 research studies on agricultural production functions carried out in the United States between the 1940s and the 1970s. The number of references from the journal *The American Journal of Agricultural Economics*, previously the *Journal of Farm Economics*, suggests that it is an important vehicle for the dissemination of such experiments.

The above casuistics supplemented by the results of multiple empirical research, has led to the proposal of a hypothetical general curve of how herbaceous crops respond to the application of nutritive substances. This curve is originated by the mere juxtaposition of countless field experiments. See figure 48, in which the top part shows how the crop reacts to larger amount of fertiliser, while the bottom shows the average (continuous line) and marginal product (dotted line) curves.

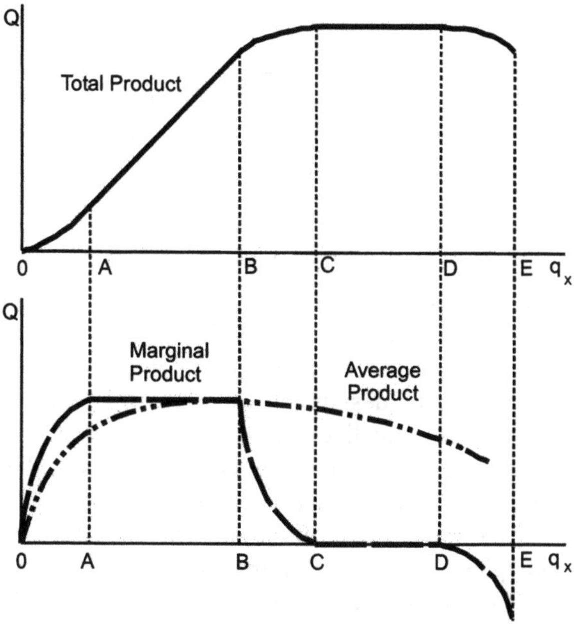

Fig. 48. Fertiliser response curve

Assuming that the fertiliser element is initially present in a limited amount, the response curve could be divided as per the sections shown in the figure. The first three stretches are controversial. Some authors consider a lineal relationship for the whole of the *0C* pathway satisfactory. Others, to the contrary, opt for a non-lineal relationship: a curve that rises at a decreasing rate. That would imply the assumption that there is a degree of interaction between the different fertilising substances. The figure shows a mixed example: interval *0A* in which the product grows more than proportionately, followed by a lineal response (section *AB*) and, lastly, section *BC* in which the product grows at an ever slower pace.[11]

[11] It must be noted that when looking at plots that have never been fertilised, or have only been so insufficiently, an exponential response curve best fits the

There is a general agreement that the response curves rapidly reach a plateau: the increase in output falls to zero (interval *CD*). The slope of the section leading up to this horizontal interval indicates the efficacy of absorption. The steepness of the slope will lessen if some physical, chemical and/or biological circumstances prevent manure assimilation, meaning that the fertiliser would be wasted. Thus, the specific dose of fertiliser that would lead to such a situation would depend on the type of soil and the farming technique used. To all events, the most relevant aspect of the appearance of this flat interval is the fact that there are maximum doses. Besides that, there is also an absence of any significant degree of substitution amongst nutrients, which can not be confused with interaction between them (Ackello-Ogutu et al. 1985). Finally, section DE represents the unanimous agreement of the agronomists that extremely high doses, unimaginable in any rational fertilising method, end up negatively affecting the yield of the crops.

As can be seen, the lower part of the figure shows the Average and Marginal Product curves. Note that curves are far more complex than the ones proposed by the microeconomists, which try to root production functions in the fieldwork experience of agricultural engineers.

It must be added that the design and estimation of production functions is also a common practice in livestock farming. Such research seeks, on the one hand, to reduce the amount of excess nutrients in the feed rations, the composition of which will vary according to the stage of growth of the animal. On the other hand, the aim is to adjust the outlay on feed to comply with the principles of economic optimisation (Boland et al. 1999).

Unfortunately the confluence of engineering and economic theory does not take into account that the research work by the agronomists:

1. Operates on a different epistemological plane to that of economic analysis.
2. A careful look at the empirical results of agricultural production will not confirm certain economic postulates. Such would be the case, for example, of the substitutability of the factors.

---

data. However, one should not loose sight of the fact that the on-going use of nitrogen fertilisers depletes a large part of the bacterial colonies that fix the nitrogen in the soil. Moreover, intensive farming facilitates soil erosion and thus affects the capacity of the land to retain nutrients. Therefore to maintain a high per hectare yield, large amounts of fertiliser have to be applied in a sustained way, even though this may seem excessive. In such a case, the path of the response curve grows at a decreasing rate.

With respect to the first point, it must be said that the theoretical goals of engineering are limited. The response curves attempt to provide an appropriate model, to the extent that they prove useful, with respect to the fragment of reality studied. This pragmatism is characteristic of the technological disciplines. Indeed, the only aim is to detect the existence of a systematic relationship between the variables chosen. After that, such statistical regularities will have to be accounted for with the help of the accumulated knowledge about the underlying physical, chemical or biological processes. This practice is very distant from production functions, which are of a hybrid nature, i.e., both technical and economic (Paris and Knapp 1989). Indeed, after establishing certain general assumptions regarding the process to be studied (such as single output or instantaneous production), and after having selected the factors that will explicitly appear in the function, different mathematical formats are tested, and the one that best supports the verification of the optimisation rules is chosen. The scope of the research is to predict the impact on the output of different variations in the amounts of the inputs considered. The estimated function is then algebraically manipulated, together with any additional information if required, to find its maximum technical limits, the marginal productivities of the factors, etc. and finally the cost functions.

On the other hand, response curves are merely an empirical instrument to calculate the right dose of fertiliser required per unit of surface area, given a plot of land with its own specific edaphologic and climatic features, framed with some agricultural methods of farming. Thereafter, the farmer will merely multiply this unit dose by the appropriate factor according to the size of the plot to be treated. Although comparing of the amounts of fertiliser for the same type of soil and nutrient has shown that there are doses at which the product will increase less than proportionately, or will indeed even fall, the everyday practice of the farmer does not seem to be affected by this. In effect, the aim of research into agricultural engineering is to detect such anomalies and inform the farmer of the dose of fertiliser that will maximise the amount of product obtained per unit of surface area.

The differences between the response curves and production functions are by no means trivial, since they reflect a profound discrepancy in terms of the nature of the research: the former represent a timid first step in a far broader study, while the latter are an authoritative prescription, designed to provide a comprehensive and definitive assessment of the efficiency of a process. Engineering experiences do not tend to *verify* economic postulates but *use* appropriate tools for pragmatic purposes. Therefore, response curves are inclined to be of little formal complexity. In engineering studies, formal simplicity is preferred as, with no more than the empirical evidence, the most complex models are groundless (Bunge 1998).

Moreover it must be remembered that the greatest concern of the agronomist is the design of the field sample. The cost of the experimental programme, the need for it to be representative and the ease of follow-up all require a thorough preliminary study of the sample. Several parcels of the same size are prepared, often adjacent to each other, in order to attempt to determine the right doses of specific material factors (fertiliser, pesticide, water, and so on) by estimating response curves. It must be noted that after a huge number of trials, such functional forms of testing have been refined. Such functions will often merely permit an approximation of the order of magnitude of the appropriate fertiliser. No universal algebraic relationship has been found. They therefore represent no more than a supplementary instrument which, with the help of plant physiology, improves our knowledge of the absorbtion process of fertilisers by plants. This is essential in order to determine the specific dose required. Needless to say that such knowledge may only be applied in other similar agricultural and climatic contexts.

Prior to closing this brief review of the methodological and theoretical problems of the production function, the postulate of the separability of inputs may be examined. Whether it is in agricultural or industrial processes, the factors tend to present a very high degree of complementarity. Here the notion of the *limitational* factor should be restored: the input whose increase is a necessary, though not in itself sufficient, condition for the rise of the amount of product (Georgescu-Roegen 1966). In other words, its impact on the level of production will depend on the increase of the amounts used of at least one of the other factors. This acknowledges the complementarity that characterises all production processes: tasks are an inextricably combination of specific machines and work, along with certain flows. For each task, the technical ratio of the worker and the system of machinery that he/she operates or supervises is only satisfactory within certain very narrow limits. To alter the worker/machine ratio, fixing one of the factors and adjusting the other, would not help increase the output and could, indeed, be absurd from an operational point of view.[12] The difficulty of separating factors means that their respective marginal productivities could not be defined (Pasinetti 1980).

Another problem is related to the presumed law of diminishing returns of the factors. Using other words, this postulate was established by Ricardo

[12] Care should be taken not to confusing complementarity and functional interaction. While the former refers to the permanent cooperation of production inputs, the latter speaks of the reciprocal implications of the different environments involved in the management of a production process (or of a firm: marketing, finance and so on).

with regard to the land rent question (or the income distribution). Observing what happened in farming in his time, Ricardo believed that the growth of the productive capacity of the agrarian sector would progressively fall further and further: sooner or later, the output from agriculture would shrink below the needs of the growing (urban) population and the requirements of industry.[13]

The modern version of this presumed law should not be confused with the hypothesis of diminishing returns of *factors* as advocated by Ricardo. Indeed, in accordance with him, this postulate is acceptable if, and only if, the volume of output changes when the amount of a given factor varies, provided that quantities of the other factors remain invariant and yield the same productive services. This situation is associated with activities involving a non-reproducible invariant factor, such as agricultural land, which presents a different productive response according to the intensity of use (in a short term horizon). Moreover, the hypothesis enunciated by Ricardo may also be understood as decreasing returns of *scale*: the different productive quality (fertility) of land yields to decreasing returns of scale (or cultivated surface). In both cases, the absence of technical change and the non-reproducibility of the factor (land) are strict conditions for the appearance of diminishing returns.

It should be remembered that Ricardo did not use terms such as decreasing returns of factors or scale. These concepts did not exist in his time. Ricardo's analysis of decreasing returns is the outcome of the economic process of ordering fertility, as well as the declining response of land to the growing intensity of cropping (intensive and extensive margin). It should be added that weather and relative prices may modify the crops sown and, therefore, the fertility arrangement and factor returns could change as well.

As stated, Ricardo's hypothesis is critically based on the implicit assumption that there are no technical changes. Regrettably, Ricardo did not witness the revolution in agricultural productivity given that it started in the middle of the 19th century. Afterwards, it has been difficult to sustain the validity of the presumed law of decreasing returns for a lack of sufficient empirical support.

It should be noted that flexible behaviour is a typical feature of living beings (soil, plants, and animals). Moreover, the intensity of use especially affects the performance of non-reproducible goods (they can show congestion problems, for example). In contrast, machines are reproducible goods and have a very lineal behaviour over a very broad range of use intensity. From that, leaving technical changes aside, the complementary relation-

---

13 This trend is only workable for the primary sector as a whole, not for a specific product (given that certain crops may be extended at the expense of others).

ship and lineal responses in production are predominant in the short-term approach.

Criticism about the theory of diminishing returns does not mean advocating constant returns as the most common situation in reality. All the discussion in preceding chapters highlights the importance of the increasing performance of the line processes. Firstly, Adam Smith made it clear that the technical division of labour would increase the skill of the worker and so reduce the execution time of production operations and thus also the idle periods between tasks. This improvement of the productivity of manual labour does not, in principle, require any great change to the tools used. Although the amount of inflow will grow in proportion to the increase in output produced, the specialisation of the worker reduces the man-hour requirement per output unit as the flow expands. To all events, the increasing technical division of work also stimulates improvements in the design of the tools and technological innovation (the invention of more efficient specific machinery). Secondly, Charles Babbage and John Rae insisted on the ability of the Factory System to reduce the idle times of the funds, a fundamental goal given the large number of fixed funds present in industrial processes. Marx also stressed the importance of the so-called increasing returns that could be achieved by combining technical improvements to the equipment (faster operating speed or greater size) with improvements to the organization of work (training, coordination of workers, etc.). It must be remembered that for the Classical school economists the price of merchandise is determined by the conditions of production and not by any assumed functional link between the returns and the amount of output generated.

Thirdly, Sraffa (1926, 1986) distinguished between increasing returns with a constant factor and increasing returns of scale. The former only appear when the input which is constant (i.e., its amount cannot be changed at will) is *indivisible*. In that situation, an increase in the quantities of the variable inputs used means achieving increasingly more efficient proportions with respect to the indivisible input. From that output will grow until attaining an optimum balance of the two types of inputs. However, the constant factor may be divisible, as is the case with land, giving rise to the increasing returns. Indeed, the amount of land required may be adjusted according to the availability of the variable factors.

Increasing returns of scale are associated with an increase in all inputs. They can only occur when no constant factor is present. With divisible inputs, increasing output will always be related to changes in the scale of production. If such increased production benefits just one single firm, it will come to hold a dominant position in the market.

Returning to the rationale that supports the presumed law of diminishing returns, reference tends to be made to the of diseconomies of scale that arise as firm increases in size. An in-depth look at this question shows that the limiting factor is the *rate* of growth and not the absolute firm size. In effect, it becomes progressively more and more difficult to sustain a high rate of expansion. The gradual exhausting of the more profitable investment projects reduces the rate of returns for the company and, at the same time, its growth.

Finally, before closing this chapter, it should be recalled that the production function model groups the elements that participate in the process in a few qualitative factors (often capital and labour). As is well known, such aggregation presents serious problems, especially in the case of capital:

> The goods (or services of long-lived goods) can be expressed as quantities of labour time performed at various dates in the past when the goods were produced. But we can no more sum this labour time in terms of man-hours than we can sum apples or bilberries in terms of fruit. To sum the cost of expenditures of labour time at various dates it is necessary to accumulate them from the time when they were performed till to-day at some rate of interest, subtracting from this cost the yield with which they must be credited for value that they have produced meanwhile (Robinson 1955: 68).

This aggregation of goods cannot be measured in physical terms, that is, without taking into account the rate of return or, in other words, the changes in the income distribution. Moreover, on occasion it is even hard to understand certain specific results. For example,

> the interest in calculating the elasticity of production with respect to the capacity of management is notably limited by the fact that it is a variable that cannot be measured directly and it is very hard to imagine what an x% increase therein would truly represent (Faudry 1974: 712-713).

If the partial equilibrium analysis considers each capital item separately and account for it, this will provide a truer representation of reality, although it must be said that it would be hard to meet with the requirements of the well-behaved economic models.

4.2 The process in line and the production function

The dissatisfaction caused by the methodological and theoretical weaknesses of the production function justifies developing an alternative approach. Circumscribed by partial equilibrium, with its pros and cons, the funds and flows model allows a better approximation to production processes following the generic concept of the production function. As has been explained, any complete set of the production processes may be described by the functional,

$$\overset{T}{\underset{0}{Q}}(t) = \Psi\left[\overset{T}{\underset{0}{F}}(t);\ \overset{T}{\underset{0}{U}}(t)\right]$$

along with the tables of productive elements which are associated with it. Nonetheless, in the case of a line process, this functional presents a far simpler configuration: it becomes a function-point which is very close to the conventional production function. It must be pointed out that this resemblance has a merely descriptive scope.

The duration of the line process is defined as the interval [0, *J*], extending from the moment the working day starts to the instant it ends.[14] Graphically, the working day is equivalent to the time lapse between the two central vertical lines in figure 49. This distance is a multiple value of c^*, given that it contains consecutive processes separated by this lag. As can be observed, when the working day starts, some units of the output are in process, i.e., they are activated at different stages of the elementary process. The same is true at the end of the working day. This is repeated every day. It may therefore be considered as a stabilised line process. For the sake of simplicity, transitory imbalances caused by changes in the rate of activity are excluded.

The boundaries of a line process are represented by the processes that are active at a given instant of time. Observing the significant part of the figure from bottom to top, the profile of the internal organisation of the elementary process can be detected. Thus, the frontiers of a stabilised line and elementary processes enclose the same content.

Under normal conditions, the continuity of the process in terms of time can be represented in a highly compact form. Given a particular production technique, it is sufficient to indicate the amounts of the flows and total service time of the funds per working day (the unit of time considered in this case). Due to the regularity of the production process, the incidence of the corresponding funds and flows may be reduced to simple points. Thus gives an extremely brief formal expression of the process.

14 This time unit has been adopted as it is the most common. Any other (hour, week, month, etc.) would have served equally well without affecting the argument. It is merely a question of convention.

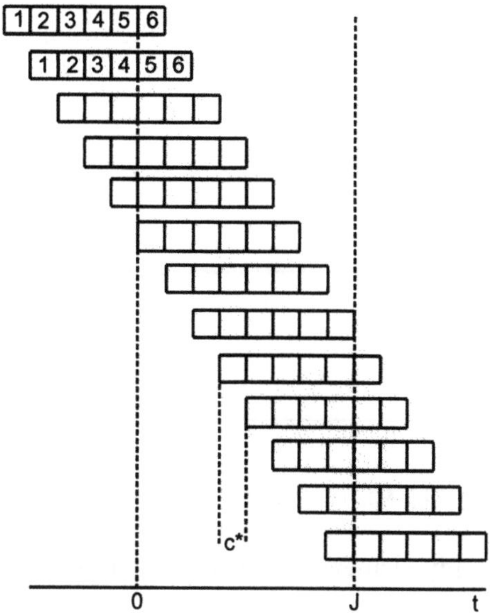

Fig. 49. A balanced line process

Let $U^L{}_j(t)$ be the function associated with each fund element present in a line process (denoted by the superscripted L). If it is valid,

1. Commensurability is assumed, that is, the duration is the same for each of the different phases (tasks).
2. There is no tolerance with respect to the lapses of inactivity of the jth fund element, that is, e_j=0,

Then it may be stated that (Tani 1986: 229):

$$U_j^L(t) = \vartheta_j = \frac{1}{c}\sum_s d_{js}$$

The expression may be directly interpreted: the function $U^L{}_j(t)$ is a *constant* for any t. Its value is the number of jth fund units employed in the process. This magnitude has a double determination,

1. The size of the lag between consecutive processes: the lower c is with respect to c^* or the shorter the lag between successive processes, the greater the number of fund units required.
2. The sum of the intervals of activation of the fund in the elementary process: the greater the said lapse of time, the more fund units need to be employed, assuming that T is held constant.

As the fund units are active throughout the whole interval [0, J] the graphic representation of the function $U^L{}_j\,(t)$ would be as shown in figure 50.

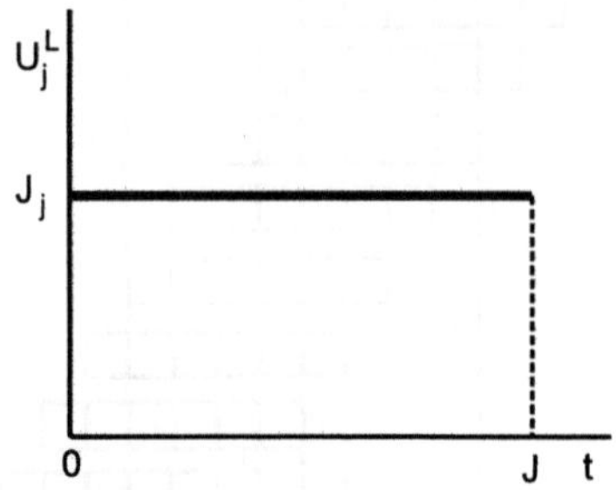

Fig. 50. The constant function $U^L{}_j\,(t)$

On the other hand, any elementary process complies with,

$$\int_0^T U_j^L(t)dt = S_j(T) = \sum_s d_{js}$$

$S_j(T)$ being the accumulative function of the services rendered by fund j.

Incorporating this into the prior expression,

$$U_j^L(t) = \vartheta_j = \frac{1}{c}\int_0^T U_j^L(t)dt = \frac{1}{c}S_j(T)$$

is finally obtained. In the general case, with $e_j \neq 0$ (j=1, ..., m),

$$\vartheta_j \geq \frac{1}{c}S_j(T)$$

where the inequality corresponds to the case of non-commensurability.

The value of the corresponding function $S^L{}_j\,(t)$ is $S^L{}_j(t) = \vartheta_j \cdot t$. The total service rendered by the jth fund element within the line process is the number of units of this fund multiplied by the duration of the lapse of time during which they are active. This number corresponds to the maximum value on the y-axis of $S^L{}_j(J)$. Indeed, if there is commensurability, i.e., e_j=0, then:

$$S_j^L(t) = \frac{1}{c}S_j(T)\cdot t = \frac{1}{\delta}S_j(T)\cdot t$$

The graphic representation of $S^L{}_j(t)$ is shown in figure 51.

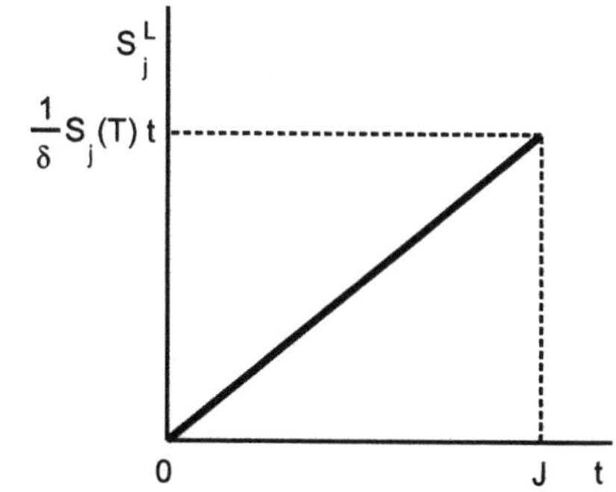

Fig. 51. The function $S^L_j(t)$

In substantive terms, the function $S^L_j(t)$ indicates the total number of service hours rendered in a line process by all the units of a given fund during the working day.

In order to determine the pathway of the functions corresponding to the flow elements, it should return to the idea stated above: their amounts may be calculated by reading vertically the process deployed in-line at intervals of length c. If the amount of each flow that enters in each interval c of an elementary process is denoted by $F_i(T)$ (i=1, 2, ..., I) to obtain then the F_i^L function, the line process must firstly be broken down into segments of time with duration of c, and secondly all the accumulative functions obtained must be joined until the whole working day is covered.

The graph of the functions F_i, defined for periods of c, does not decrease. A hypothetical example is shown in figure 52. It corresponds to a flow initially entering at a decreasing rate which later becomes constant. There are, of course, other possibilities. Thus, the pathway of the F_i^L function may be uniform, as shown in figure 53a, when the flow enters the process at a constant rate, or it may follow a twisting pathway, as in figure 53b. In this case, the flow first enters at an increasing rate and later at a decreasing one.

To all events, the F_i^L functions are developed around an axis centred on the origin of the coordinates. Then, in taking pathway F_i^L as uniform,

$$F_i^L(t) = \frac{1}{c} F_i(T) \cdot t$$

can be set up. Ceteris paribus, during the course of a working day the smaller (greater) the lag between consecutive processes is, the greater (lesser) the amount of flow. If the value c lessens, a larger number of processes will be executed per unit of time and, therefore, a greater amount of flows will be consumed.

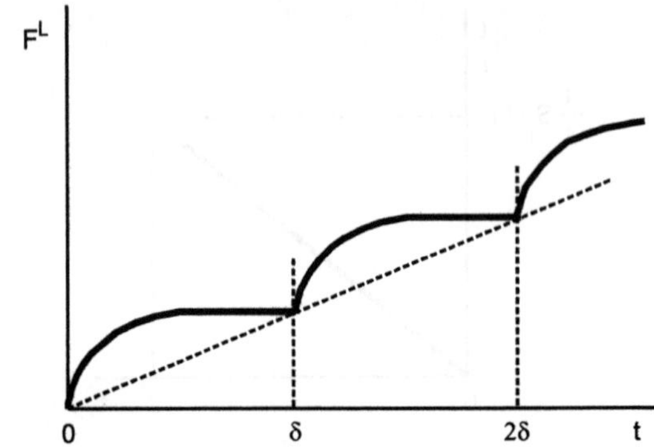

Fig. 52. Hypothetical function of flow F_i

The representation of the services of the funds and the consumption of the flows of a line process has shown that the two accumulated amounts are proportional to the reduction of the lag between processes and the duration of the working day.

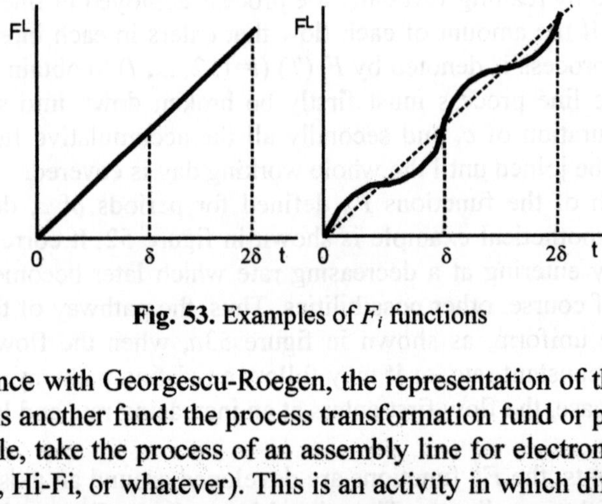

Fig. 53. Examples of F_i functions

In accordance with Georgescu-Roegen, the representation of the line process contains another fund: the process transformation fund or process-fund. For example, take the process of an assembly line for electronic consumer goods (TV, Hi-Fi, or whatever). This is an activity in which different flows of components are progressively assembled until the product is finished. Over the course of the working day, components enter at the beginning of the line and come out as finished units at the end. Seen as a whole, the impression is of being an *instant* process: the input entrance and the exit of the finished product are apparently simultaneous. However, what we are seeing are different units. Nonetheless, looked at globally, the process seems instantaneous. The output in process, distributed along the line, permits this rather special view of the production activity. These unfinished or semi-manufactured goods constitute the so-called process-fund.

Its basic feature is to reproduce, at each different moment of time, the complete structure of the elementary process. This fund element coincides with the region of stable configuration of the line, a region characterised by a continuous movement of units of output in process. Despite the fact there are some doubts about the process-fund because of its passive role in the process, it reflects the continuity of the process in line. The inclusion of the process-fund means giving completeness to its representation. It is also essential in order to understand why the process in line can be represented as if it were an instantaneous production process. In short, eliminating it from the production function would result in a loss of important information regarding the process in line.

The functions that represent the process-fund are shown in figure 54. At the top of the figure there is a constant function $U^L{}_{m+1}(t)$, while $S^L{}_{m+1}(t)$, lineal within the interval [0, J], is represented beneath the associated function.

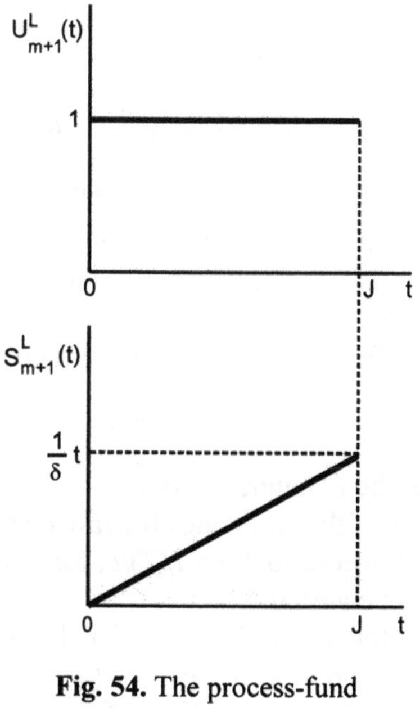

Fig. 54. The process-fund

For a single output line, the ordinate value of function $U^L{}_{m+1}(t)$ will always be 1. Indeed, along this line, an output unit can always be identified at each of its different manufacturing phases (Georgescu-Roegen 1996: 306). For this reason, the scale of the fund of semi-manufactured goods is the unit.

On the other hand, the total number of process fund units will be obtained by dividing the length of the considered unit of time by the lag between processes. That is,

$$S^{L}{}_{m+1}(t) = 1/c\ t$$

If the lag is shortened, a greater number of process-fund units will be required. In the case of extending the working day, the process-fund will increase by $\Delta t/c$ units. To this extent, the modification of the process-fund will be as expected to result from by technical innovations in flows and/or funds. Its composition will moreover change due to the productive particularities of different plants (Landesmann 1986: 298).

Having reviewed the representation of the elements of the line process, it may now be fully symbolised. To such ends, flows are sorted into output flow (Q), natural resources (R), raw material (I), by-products (W), and maintenance (M), and for the funds, capital assets (K), the process-fund (K_{m+1}) and the human work (H). The following is the expression for a line process (adapted from Georgescu-Roegen 1972: 283ff., 1976: 67; Ziliotti, 1979: 641ff.):

$$q = \varphi\,(r, i, w, m; k, k_{m+1}, h)$$

the components, scalars or vectors of which may be read as follows:

1. q represents the main output obtained at the end of an elementary process (its duration is T). The amount of product obtained per unit of time (one hour, for example) is $1/c$.
2. r is the vector of the flow rate for the different natural resources per unit of time. Thus, $r=(r_1, ..., r_n)$, being each $r_i = R_i(T)\cdot\frac{1}{c}$, for $i=1, ..., n$, refers to the flow of the ith natural resource.
3. i indicates the vector of the flow rates for raw materials per unit of time. That is, $i=(i_1, ..., i_n)$, where each $i_i = I_i(T)/c$, for $i=1, ..., n$, corresponds to the flow of the ith raw material.
4. w represents the vector of the flow rate for the by-products per unit of time. As with the two previous cases, $w=(w_1, ..., w_n)$, and $w_i = W_i(T)/c$, for $i=1, ..., n$, which represents the flow of the ith by-product.
5. m refers to the vector of the flow rate of material required for maintenance operations per unit of time. Here, $m=(m_1, ..., m_n)$, where the $m_i = M_i(T)/c$, for $i=1, ..., n$, corresponds to the flow of the ith.
6. k represents the vector $k=(k_1, ..., k_m)$, the components of which indicate the *units* of the different fixed capital assets active in the process at a given moment. Thus, $k_j \geq S_j(T)/c$, for $j=1, ..., m$.

7. k_{m+1} refers to the unity of the process-fund, always present in the production process.
8. h indicates the vector $h=(h_1, ..., h_m)$, formed by the *number* of workers of each different kind active at a given moment. For $j=1, ..., m$, the result is that $h_j \geq H_j(T)/c$, where $H_j(T)$ is the accumulated function of the services of the jth fund in the elementary process.[15]

This may also be written as follows,

$$Q = \vartheta(R, I, W, M; K, K_{m+1}, H; \tau)$$

This equation contains the amounts of the flow elements and fund services, τ being the duration of the working day or any other unit of time. Given that $R=r\cdot\tau$, $I=i\cdot\tau$, $H=h\cdot\tau$, and so on, the function ϑ is homogeneous to the first degree with respect to all the variables. Hence,

$$\tau\varphi(r, i, w, m; k, \mathrm{k}_{m+1}, h) \equiv \vartheta(R, I, W, M; L, K, K_{m+1}, H; \tau)$$

from which,

$$\varphi = \vartheta, \text{ for } \tau = 1.$$

This result

> does not mean that the factory process operates with constant returns of scale. The homogeneity [of ϑ] corresponds to the tautology that if we double the time during which a factory works, then the quantity of every flow element and the service of every fund will also double (Georgescu-Roegen 1976: 67).

The two above expressions are not functionals but functions. In the case of a line process, the functional has become a *function-point*, i.e., it is a mere set of numbers.

Despite certain superficial similarities, it must be pointed out that this representation has significant differences with respect to the conventional production functions. The reasons are:

1. It is only legitimate for line processes in which the funds are permanently active and the flow rates are constant at all times. The elements participate in the production line at a rate which will not vary with respect to time (Ziliotti 1979: 642),
2. While the production function is based on the theory of diminishing returns, nothing is established with respect to it in the funds and flows model. At most, what is assumed is that φ and ϑ will normally be monotonic and not diminishing functions.

---

[15] In the case of k and h, the imbalances become balances if the periods of inactivity are completely eliminated (Tani 1986: 233).

3. The level of production is under the influence of c and τ. In other words, it depends both on an increase in the pace of activity and the lengthening of the working day. For this, the model may incorporate changes in line speed along with the use of different shifts.
4. One should not forget the general theoretical framework to which the concepts employed pertain. For example, it does not speak of factors in general, but of flow and fund elements, which are not at all the same thing.

In short, the function $q = \varphi(r, i, w, m; k, k_{m+1}, h)$ possesses a composition similar to a list of ingredients: it shows what a plant is potentially capable of, not what it actually does. The equation describes,

> the process in the same manner in which the inscription "40 watts, 110 volts" on an electric bulb, or "B. S. in Chemical Engineering" on a diploma, describe the bulb or the engineer. Neither description informs us how long the bulb burnt yesterday or how many hours the engineer worked last week. Similarly, [the functional] may tell us that a man with a 100 hp tractor, which uses three gallons of gasoline and one quart of oil per hour, can plough two acres per hour (Georgescu-Roegen 1971: 241-2, inverted commas in the original).

5 The fund-flow model and service activities

Even though the fund-flow model was developed to represent production processes with tangible outputs, there is nothing to prevent its application being extended to other forms of activity which generate added value. This chapter offers a preliminary investigation of the application of the fund-flow model to service activities, with special reference to transport operations and the development of information assets. The rationale behind such an extension is quite simple: a pure service does not exist. In rendering it, the direct and indirect participation of certain funds and/or flows will be required. However, this incursion into the area of services does not aspire to be anything more than an introduction. Tertiary activities form a highly diverse and ever-changing universe. Consequently, attempting to cover their idiosyncrasies in any depth would require a study that goes far beyond the scope of this book.

5.1 Characterization of services

In developed economies, the service sector has progressively gained in importance, both relative to GDP and employment. It has become by far the predominant sector. Specifically, over the last 50 years, health related activities and care services have grown, as have education and corporate services as well. In some countries, such activities employ, in absolute terms, almost as many workers as the primary and industrial sectors together. Although the rate of growth of these types of services may slow in the future, the complexity of the production and marketing processes, as well as the citizen pressing for more and more educational, cultural, care and health services, ensure their expansion.

Comparing professional categories, the increase in tasks that can be considered tertiary has been spectacular. Despite this, the general growth in white-collar workers should not conceal the swift increase in the number of blue-collar workers. These are employed in service activities, such as the retail trade, transport or private security. When accounting for this growth, it is important to drawn attention to the outsourcing phenomenon (from staff selection and accountancy to maintenance and cleaning). Tasks previously performed by industrial workers are now subcontracted out-of-house to third firms. Their employees, from the statistical point of view, are classified as pertaining to the service sector. It must be said that out-sourcing is a practice used by firms in both the manufacturing and service sectors.

It is not easy to define services. It covers such a vast array of activities that some authors claim that the use of a single label would be of no relevance whatsoever. It is then enough to identify services by contrast with to industry and agriculture, but it is by no means a satisfactory solution. Indeed, this definition pays no heed to the differential sectoral traits. Fortunately, there are many authors who consider services as an area of independent economic activity. For this reason, it deserves to be studied as an independent category. For example, Smith, Say and Marshall characterised services as activities producing an intangible output. Or, in other words, services offer an output that vanishes at the very instant of its use. A first step to define services is to conceive of them as a process of exchange between economic agents involving added value. This approach underlines the fact that, presumably, services may neither be stored nor transported (Gadrey 2000: 370).

In general, it is considered that the output of any process of transformation is tangible. It means that it will have a physical materiality that may be precisely situated in time and space. Note that the concept of economic process may be extended to cover the changing of the *state* of a given reality, because an agent has been rendering a particular service. This enlarged definition of economic process does not cause any irremediable conceptual damage. Firstly, observable changes of state may include a huge variety of events: the improvement of a person's health after a surgical operation, the enhanced efficiency of a machine after maintenance and repair or the accumulation of knowledge through access to new information. Secondly, many services are provided by the direct interaction between people, but there also are self-service and anonymous (as is the case of defence) service provisions. Finally, even though services may not be generally stored, there are cases, like information, where they can be accumulated.

When speaking of services, there are two common misconceptions to avoid. The first is to confuse the act of rendering a specific service and the physical resources (funds and flows) required by this. For example, the passenger transport service could be rendered by trains, cars, etc., each of them using infrastructures of a highly diverse nature. The second is not to confuse the service provision (e.g. giving a master class) with its more or less enduring effect on the agent receiving this service (the broadening of his/her knowledge base).

A pioneering definition of the service relationship is to be found in Hill (1977). It is based on two criteria:

1. It distinguishes between the process of the service and its result, i.e., the change of condition or state of a specific reality.
2. It specifies the agents involved and the reality modified (the subject of the activity).

Given the above criteria, *service* can be understood *as an operation intended to modify the condition or state of a certain reality (thing, person or organisation –hereinafter called C), possessed or used by an individual (he or she who demands or consumes the service - B), and performed by a given agent (a fund element)*. This agent (hereinafter called *A*) performs the service at the request or demand of *B*, who normally assumes a commitment to make a payment in return. The juridical nature of agent *A* may be very diverse.

The economic operation of providing a service may not be physically separated from the entity *C*. This entity may consist of objects that have been modified in a given way (for example, moved in space), or people or organisations that have been altered in some of their physical, intellectual, functional or other dimensions. Elements *B* and *C* may be one and the same as, for example, the same person in the self-service case, or separate entities. In this case, *B* supposedly has some rights (or decision making capacities) over *C*. Of course, the act of service providing is the specific intervention of *A* (a fund) with respect to *C*. Agent *A* may simply convey information or may execute an activity (with the aid of a given quantity of other funds and/or flows, and taking up a certain amount of time). This intervention may consist of making permanent or transitory, reversible or otherwise, changes to *C*. Moreover, in rendering services it can use greatly varying degrees of diligence and attention, along with very different procedures and technologies. Consequently, the resulting level of quality of a given service, using an appropriate indicator, may vary greatly.

As seen, the relationship involved in the provision of services has three poles: two economic units (*A* and *B*) and one modified entity (*C*). The decision making process and finance of services are elements of the relationship between *A* and *B*, while any discussion regarding their pertinent public or private provision is associated with the legal nature of *A*.

There is no doubt that *C* is the key element of service definition. To be more specific, the task of the service provider *A* to the benefit of *B* may be any of the following:

1. *A* ensures the execution of a given service, such as transport or repair on tangible objects possessed or controlled by *B*.

2. *A* manages or transfers coded or tacit information (including money) on behalf of *B*.
3. *A* alters the physical or mental condition of *B* in some way.
4. *A* analyses and/or transforms organisations, or certain aspects of these (configuration, know-how and knowledge) often by mutual agreement with *B* (whether he/she is a member of organisation *C* or not).

The conceptual triangle proposed below is based on the perception of a service as a procedure which leads to a change in the condition or state of a person, organisation or thing. All of that is the result of an agreement established between the agent who possesses the reality to be intervened upon (which may eventually be the individual them-self), and the other agent. This latter has been commissioned by the former to make the desired change/s. The output of the operation is therefore a condition or state change.[1] The relationship between these elements is summarised in figure 55.

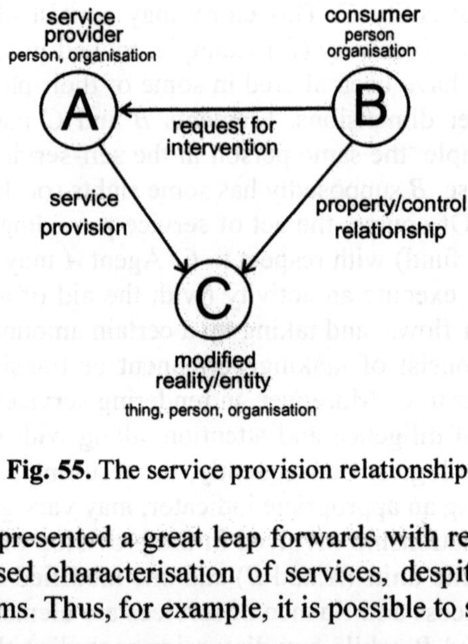

Fig. 55. The service provision relationship

This definition represented a great leap forwards with respect to any other previously proposed characterisation of services, despite harbouring certain minor problems. Thus, for example, it is possible to see the following:

1. It should be pointed out that the companies working in service sectors, such as telecommunications, hostelry, catering or retail trade, offer compound services. For example, a hotel offers rooms for overnight stays, rooms for social events, information to the client with respect to

[1] Although not stated in the definition, it is assumed that services are supplied in a market economy.

questions of his/her interest and supplementary services, such as the restaurant or leisure activities. Each of the aforementioned services modifies the condition or state of the client in one way or another. Firms working in the hotel and catering sector are well aware that the service of accommodation is a complex and subtle task. These features are used as a mark of quality and distinction to obtain advantages with respect to the competition. This is also the case of the retail trade which brings goods to the client coupled with information about these (Gadrey 2000). In short, it should be remembered that one single organisation may offer multiple services, either simultaneously or successively. All these services are capable of altering one or more dimensions of the condition or state of the customer.

2. From the conceptual point of view, the service understood as the output of a process which goal is to provide it should not be confused with the productive services performed by funds in a manufacturing process. Although both represent a change in the state of the object intervened on (the customer or current output), from an analytical point of view it must distinguish between the act of rendering a service as merchandise, and the execution of productive operations within a given process by the human work and fixed funds.

This last point is important since it makes a distinction between two types of merchandises:

1. The work performed by a fund: in purchasing work the generic capacity is acquired to render certain services over a given interval of time. The specific use of those will depend on the process organisation, be it in the agricultural, manufacturing or tertiary sector.
2. The rendering of a given service: in purchasing a service an activity is acquired which is ready for sale and carried out with the cooperation of funds and flows in accordance with a pre-established plan. With a more or less known format and conditions, the value of a service will depend on the amount of the direct or indirect work involved.

In the first of the two cases, the economic operation is the purchase of services of human work. In the second, it is the purchase of services as defined above.

Later, Hill refined his earlier definition, clarifying the fact that services are not entities capable of having an existence of their own:

A good is: (1) a (tangible or intangible) entity that exists independently of its producer and its consumer; (2) an entity to which ownership rights (private or public) (...) can be assigned and that can therefore be resold by its owner. A service: (1) is not an entity; (2) requires a relationship to exist between the person seeking a service and the service provider (request for intervention); (3) concerns an entity C (individual, good, material system) owned by the person requesting the service; (4) has as output S a change in the condition or modification of the state of this entity C. No specific ownership rights can be assigned to this output, so there is no possibility of S being resold independently of C (Hill quoted in Gadrey 2000: 378).

Service *S* may not be separated from *C*: it cannot be resold as a separate entity. This lack of independence of services avoids property rights being granted to these.

It has not been easy to define services. This is firstly due to the complexity of the economic relationship involved in them. Secondly, services have a great variety. To cover them all would mean painstakingly adding an ever-rising number of elements to the definition. For this reason, the characterisation proposed above becomes plain if the following two types of service are considered (Gadrey 2000: 382-4):

1. The first one is when organisation *A*, which owns or controls resources and/or a human capacities, sells (or offers without payment in the case of non-market services) the right to the temporary use of the said resources or capacities to an agent *B*. This temporary use will bring about effects deemed beneficial by and on agent *B* himself, or on goods that he owns or for which she/he is responsible. The most obvious example is the rental firms (vehicles, business premises or housing). Another case is firms that offer maintenance agreements for equipment, systems or facilities, as well as consultancy services (fiscal, legal, or whatever) and staff training services. In such situations, a firm or individual temporarily hires access to certain experiences and knowledge.
2. The second is when individuals require somebody to take care of their assets or him/herself along with other assets or people they are responsible for. This case would include domestic, personal care and surveillance services.

Without denying the validity of the differentiation proposed recently, it would be better to build a more exhaustive typology but of a manageable size. A highly developed proposal (adapted by Gadrey 2000: 384, 2003: 20-1) is based on the following two criteria:

1. The attitude adopted by *B* (the demanding agent) in taking benefits from services. These may be two types of posture: active (in rendering the service there is an interaction between the customer and the provider) or passive (the customer limits him or herself to following the provider's instructions or simply watching what the latter does).

2. The technical or human capacities used by service customers.

Mixing the two criteria, the following three way division of types of demand (demand rationales) for services is proposed (Gadrey 2000: 384-5, 2003: 20):

1. Services that appear as a response to calls for intervention or support.
2. The provision of capacities that very often require the use of complex equipment and systems.
3. All kinds of live performance mainly offered for entretaintment purposes.

The services that respond to *B*'s request for intervention or support include the care of the sick and elderly, as well as education and training for all ages and situations. Consultancy, guidance and audit services should be added, along with engineering and design and other professional services to firms, institutions or individuals. The service may be supplied publicly and/or privately. These services are extremely hard to standardise. They also have many highly subtle aspects to examine when attempting to measure their quality. Somebody thinks that the quality of these services can be measured by the temporal length of commitment which supposedly is directly related to the degree of care and exclusivity given. In some cases, service providers may supply the quality indicators, although the degree of objectivity and detail then achieved tends to be limited.

Within this category of services, the last few decades have witnessed of the expansion of outsourcing contracts. Different reasons, such as management and production cost reduction or the increasing segmentation of the markets account for this increasing demand for such intermediary services.[2] Nevertheless, there are several different situations:

1. Highly specialised services, such as banking or insurance that are farmed out.
2. External services that firms are obliged to contract for legal or institutional reasons (certification, verification of production conditions, standards, etc.).
3. Services which have to be supplied internally as the market does not offer them yet (for example, the management of the knowledge generated by a firm).
4. Services which can optionally be internally generated or outsourced. In such cases the final decision will depend on the capacities of the firm and the current market conditions.

---

2 These practices may have their own paradoxes: firms will engage IT experts while simultaneously outsourcing IT services.

The second category of services is the provision of complex technical capacities. This is the case of the transitory use by agent *B* of a series of resources, infrastructures and systems placed at his disposal by *A*. The quality of the provision of such services will depend on the correct operation of the material basis used to rendering them. Examples are hostelry and catering services, transport (by road, rail, sea and air), telecommunications (radio, TV, telephone, Internet) and trade distribution (wholesale or retail, by traditional stores or by large shopping centres). Such services tend to require large-scale investment in facilities and capital equipment. These services sometimes make up complex networks with a lot of room for productivity enhancement as well as for potential problems of compatibility. The setting up of large-scale infrastructure, such as the telegraph or railway networks, bears a cost with no comparative proportion to the size of the initial market. This casts doubts on its profitability and discourages private investment. In such context, public regulations or corporations tend to be established. This type of service is measured by the number of accesses to them or by the amount of space and/or time occupied: number of calls, average length of calls, tons-kilometre, number of overnight stays in a hotel, the number of meals served per day and so on.

The last category of services is the supply of capacities (human or animal) and the exploitation of physical phenomena for entertainment. It includes all kinds of shows or other cultural manifestations. They are aimed at a public whose main mission is to watch them. In terms of the conceptual scheme proposed, the agent *A* organises exhibitions of human creation that agent *B* watches. Notwithstanding, in activities such as tourism in nature reserves and the like, the activity of agent *A* is to facilitate good access and observation conditions. When dealing with human creations, the service tends to be a pre-packaged sequence of performances, i.e. it is an ordered set of actions following a previously established pattern. The aim is to satisfy the requirements of the audience watching. Indeed, a play at the theatre, a concert, a visit to a museum or an art show tends to follow a customary format with sufficient flexibility to be able to admit originality. The interest awakened by the use of such services may be measured, although always very imperfectly, by the size of the audience, the sale of tickets, references in the press and so on.[3]

The classification given above includes a brief reference to the criteria for the measurement of the quantity and quality of the service supplied.

---

3 Despite the division proposed and the examples given for each case, it must be acknowledged that there are services possessing mixed features. Such is the case of the postal services which have traits of both the first and second categories.

For example, for services requiring direct attention which may be interactive to a greater or lesser extent (as those rendered by a school teacher, an attorney or a doctor), the output may be measured in terms of the number of hours of activity, the number of interventions, etc. With respect to the indicators of quality, the debate is, and probably will stay, open. Another difficulty arises from the fact that certain services combine diverse elements of a widely varying nature. For instance, retail traders operate simultaneously on goods and information. An analysis of their efficiency should be broken down into parts.[4] To converge in a broadly acceptable compound indicator may be arduous. Very distant from the studies recently mentioned are there those which try to assess the short or long term impact of the continuously provision of certain services. Typical cases include research dealing with the relationship between health care services and life expectancy or amongst the consultancy services received and the profits made by a firm.

If a service requires no special knowledge or skill, the self-service provision becomes usual. In this case, customers are the agents who render all or part of the service. People perform foreseen operations using the tools and systems supplied by the establishment.

There are complementarities between goods and services. On the one hand, a buyer of a car will require the services of an insurance company, a car wash, maintenance services, and so on. On the other hand, it can be stressed that, without the existence of these services, it is unlikely that private transport would ever have become such a widespread phenomenon. There may also be complementarities between services: the diversification of tourism services gives rise to specialised travel agencies and vice versa. Another reality to be taken into account is the simultaneous expansion of the consumption of goods and services. For example, people own more television sets and at the same time go to the cinema more than ever before. After a film has been premiered, they buy the DVD, which means having a DVD player and a television screen. Such dynamic complementarities do not exclude some cases of service replacement by physical goods: instead of having a broken device repaired a new one is bought to replace it.

A singular aspect of certain services, such as the retail trade, is the existence of periods of inactivity. The employees have to be present in the establishment, even though the number of customers coming in may be low. If the affluence of people is greater than the capacity for attention avail-

4 For example, it may distinguish between the quantification of stock rotation, the assessment of the variety of products to hold, and the degree of personalised attention to give to the customers.

able, waiting becomes the adjustment mechanism. There are two ways to control the economic damage of a low level of activity: flexibility of working time and availability when required of employees, along with low wages. The tertiary sector is then known for part-time working, short-term contracts, etc.

Services have a limitless capacity for diversification. Whenever the conditions are right for the formation of a specific market with a solvent demand, a new service will always emerge. This new, or improved, service is characterised by the fact that it is tailored to meet the particular demands of the customer. In general, new services emerge from the following:

1. The combination of services. For example, the insurance and financial agents continuously create new products by endlessly combining and recombining legal rules and mathematical formulae.
2. The adding of complementary services. Restaurants and hotels are incorporating more and more ancillary services in order to draw in specific new customers. Something similar happens in consultancy firms that are offering new services, such as the evaluation of intangible assets or environmental impact certification.
3. The breaking down of what was previously a single service into several different ones. For example, selective waste collection.
4. Exploiting the options technology opens. In the area of telecommunications services such as FM, "pay-per-view" TV, mobile telephony, etc. have appeared.
5. The adoption of techniques to increase the efficiency of a service. This is for example the case of automatic inventory or mechanical bagging at the supermarket check out till.

Entering now into the conceptual scheme of the fund-flow model, the provision of services requires facilities and equipment along with human work, without forgetting the consumption of some inflows. Although the relative weight of each element will vary according to the kind of service, it must be pointed out that a pure service does not exist. Even if the service considered is dance classes, a teacher, premises, hi-fi devices and electricity are required. Without taken into account the indirect consumption of resources (by the teacher in the case in hand, for example), it is clear that the services provision involve certain funds and flows for a given interval of time. In dealing with services, however, there is a question that deserves clarifying: the nature of the output flow. In the case of services the output is not a physical good perfectly located in a given place and time. The output of services, as stated above, is a change in the condition or state of a given entity. These changes are observables and may be quantified, either directly or indirectly by way of different indicators. For example, the outflow of a telecommunications service would be the information traffic measured in bauds or bits per second; if the service is surgical operations, the health improvement of patients may be assessed by diverse indicators of recovery; in the case of maintenance and repair activities, the restored functionality of the machine can be gauged in appropriate units of measurement[5]; in logistics, the outflow is the displacement of supplies or people over a given distance; and, to close, if talking about educational or training services, the improvement in knowledge or skills may almost be measured on an ordinal scale. Figure 56 shows a general flowchart of the service providing process using elements drawn from the fund-flow model.

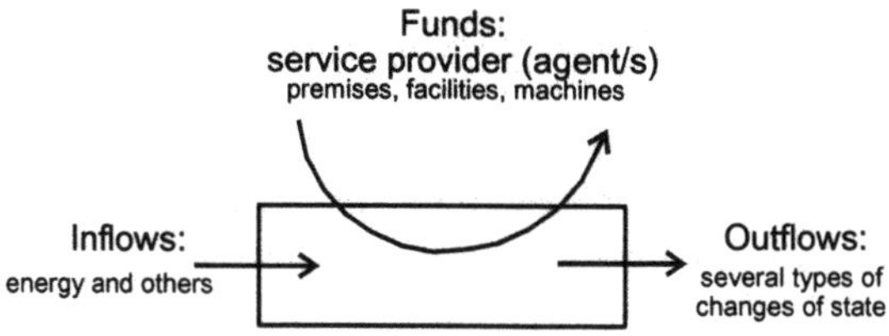

Fig. 56. The process of service supplying

5.2 Modelling of transport operations

Two types of transport operations are considered herein: the transport of people or goods over a greater or lesser distance, and the movements of funds and flows within production processes. In either case, displacements will require inflows, such as energy and a container capable of movement. This latter is a fund. The same may be said of the animals and/or people that draw/guide such a mobile platform. Transport services stand out for a peculiarity of the outflow: it is a change of position. The output is therefore identified by a given amount displaced over a certain space (in any of three dimensions). This operation obviously requires time. With regard to transport operations occurring in a production process, such operations are related to the inflows and, most especially, to the output-in-process. Indeed, as stated in chapter 1, the fragmentability of a production process

[5] The example of maintenance and repair operations is an interesting one as Georgescu-Roegen, in his attempt to simplify the representation of the production process, proposed leaving such activities aside. Nonetheless, he never returned to the question. Consequently his writings make no mention whatsoever of services.

may give rise to different phases being performed in different plants.[6] It is therefore convenient to develop a representation of the carriage of elements from one place to another.

From the viewpoint of the funds and flows model, transport operations may be described by the next simple equation representing a change of state:

$$\text{funds} \oplus \text{flows} \rightarrow [q]_{x_0-x_1}^{t_0-t_1}$$

The output of the process is the spatial displacement (x_0-x_1) requiring time (t_0-t_1) of something (q). Analytically, it is important to consider:

1. As usual, it should take account of the funds and flows involved in the process of displacement.
2. The description of the movement. This implies computing the distance travelled and the time consumed by the object displaced. This information identifies the change of state.

With respect to the second point, transport operation engineering offers two simple tools which are very useful for describing displacements: time-space diagrams and accumulative graphs (Daganzo 1997: 2-27). The configuration of time-space diagrams is shown in figure 57.

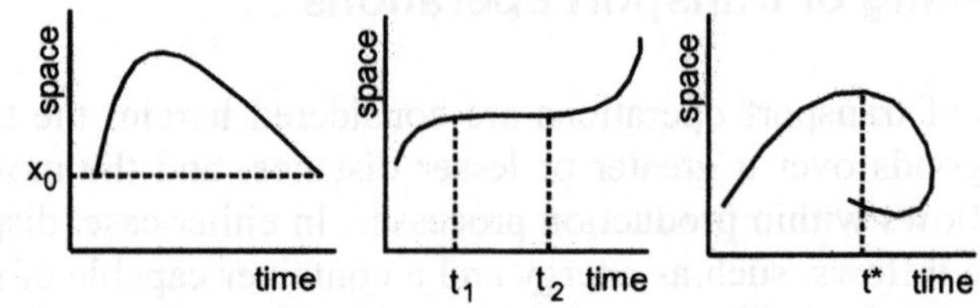

Fig. 57. Time-space diagram: pathways $X(t)$

The (t,x) space may be directly interpreted: the pathway of the left-hand figure relates to a vehicle which turns around and returns to its point of origin (x_0), while the central figure shows a stop between t_1 and t_2 and, finally, the figure on the right represents an impossible displacement because of travelling back in time. Consequently, the only valid pathways are those in which each time co-ordinate t is associated with one, and only one, x-point. It should be noted that the point of origin of the axes is taken at random.

6 As is known, this is the case of just-in-time procedures when several plants carry out the same manufacturing process.

The diagram may be used to represent any kind of movement. For example, a swift increase (reduction) in $x(t)$ denotes a fast advance (return), while a horizontal stretch indicates a stop whose duration is the length of this stretch. The straight segments show constant speed, while a curved line indicates acceleration (deceleration). Moreover, if function $X=x(t)$ is known, the speed of the transport system is $v(t)=dx(t)/dt$, whilst acceleration is given by $a(t)=d^2x(t)/dt^2$.

The configuration of a time-space diagram of a closed loop would be as shown in figure 58. In this case, four vehicles are represented distributed at regular distances, with parallel pathways in the form of closed circuits. With L being the length of the loop and $0 \leq x \leq L$ indicating the position of the vehicle in the loop, its displacement *disappears* when it reaches point $x=L$ and simultaneously reappears at $x=0$. The diagram therefore shows a saw-tooth pattern as long as the speed of the vehicle remains constant. It must be added that the loop movements of funds are usual in productive processes.

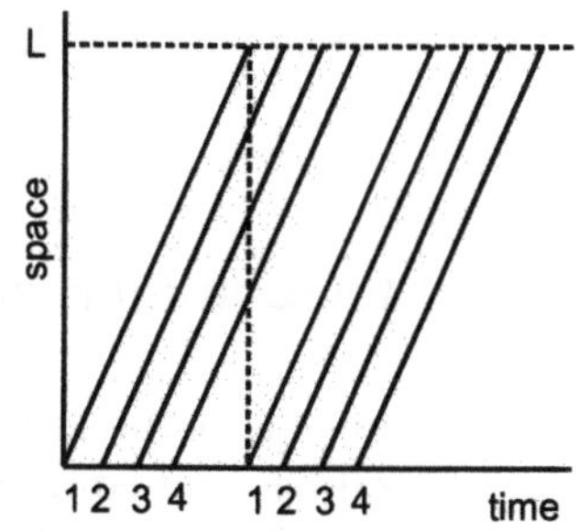

Fig. 58. Loop displacement scheme

Functions $N(t)$ indicate the accumulated number of vehicles that are in circulation at a moment t, starting from a random point in time ($t=0$). In the case of a few vehicles, the function $N(t)$ tends to be staggered. Nonetheless, if there is a very large number of circulating elements, the function may be considered continuous and apt for mathematical manipulation. It is frequently advisable to split function $N(t)$ into two: $A(t)$ and $D(t)$, accumulative input and output functions, respectively. These are shown in figure 59.

$Q(t)$ is the number of elements situated between two observations: $Q(t)=A(t)-D(t)$. The vertical separation represents the level of accumulation. On the other hand, a horizontal reading of the figure (w) indicates the time of circulation of the nth elements in the system if they circulate following the *first-in/first-out* rule, i.e., $w(N)=D^{-1}(N)-A^{-1}(N)$.

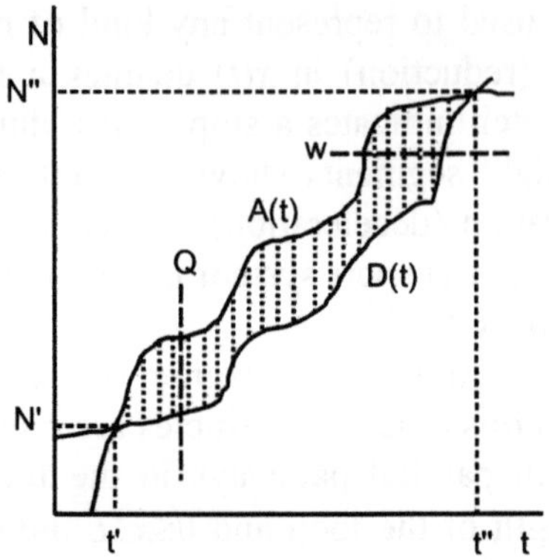

Fig. 59. Functions of accumulated number of vehicles

The region enclosed by $A(t)$, $D(t)$, t_1 and t_2 (the striped part of the figure) indicates the total system waiting time for the interval considered. The waiting time may be measured as units-hour. Finally, it must be said that such graphs could be useful when analysing the effect of restrictions on the circulation of flow element units.

5.3 Fund-flow analysis of telecommunications

Telecommunications permit distant interlocutors to exchange information through a complex network of facilities and equipment. The message may take the form of sound (words or music), images (static or animated), written documents (text or drawings) or data (figures, code, etc.). The message is transformed into a set of signals (analogue or digital) which is directed, using a transmission channel, from the sending terminal to that of the receiver. From a physical point of view, signals take the form of electric pulses, waves or bundles of light, and move through media such as copper wire, fibre-optics or electromagnetic waves. To enable communication, the signal must be sent in a code shared by both the sender and the receiver. In order to achieve a maximum number of signals over a minimum number of channels, an important technical innovation was the multiplex technology. This can assembly signals from different sources and transport them through a single channel as a composite signal.

A telecommunications network (in star, ring, bus or tree architecture) comprises three layers:

1. The ensemble of facilities and equipment that allows the signal be sent, carried and received. Elements of this infrastructure are telephone

headsets, broadcasting equipment, switchboards, cable (par, coax, fibreoptics), masts and aerials for Hertzian waves, modems, satellites, underwater cable, switching stations, and so on. All of these are fixed funds allowing the connection of the interlocutors' terminals. In effect, switching systems select and open channels through which the signals are directed and, once communication is finished, they release them for the following use.
2. Traffic management functions (command, control, measurement and optimisation) are performed by specific teams. Their capacity for work has been greatly extended with the incorporation of ICTs.
3. The services offered to the user/customer: videoconferencing, data transmission, voice calls, telephone messaging, faxes, etc.

A telecommunications network is a vast technical system. In such systems, the different components are interconnected, meaning that any change made to one of them will tend to affect the performance of the others. The most outstanding features of a telecommunications network are:

1. The weight of fixed funds investment. The costly infrastructure and equipment have a long operational life, although this life can be curtailed by obsolescence.
2. The digitisation of highly diverse contents. A digital signal is less prone by disturbances, such as background noise or interference, than an analogue one.
3. The rate of traffic flow varies greatly. However, there are predictable cycles because of their regularity. Sometimes the network becomes saturated (congested), alongside other periods of great surplus capacity. In this respect, optimising the use of the network consists of minimising the intervals between periods of saturation (Marini and Pannone 1998: 180).
4. Different technical innovations have multiplied the capacity of the fixed funds. At the same time, other improvements allow channels to be assigned to different users demanding different services. This is to take advantage of economies of scale and scope. For example, fibreoptic cables have enormously reduced the cost of transmitting telephone calls, but that of switching only to a lesser extent, while they also allow subscribers to have access to all sorts of telecommunications services.

From its earliest days until the 1980s, the telephone network was considered a natural monopoly. The importance of the economies of scale and network externalities justified centralising the telephone service in one single public operator (commonly the case in Europe), or in a handful of private operators under government regulation (USA). With liberalisation in the 1980s and 90s, telecoms could go further towards geographical exclusivity and service specialisation. International alliances led to the creation of enormous and highly diverse conglomerates that fostered expansion strategies. This firm policy was supported by the monopoly based income earned in its original market. As occurred with air transport, the need to consolidate the emerging market was at first translated into benefits for the customer. However, after achieving a dominant strategic position, these big companies returned to customary oligopolistic practices.

From a technical point of view, since the 1970s switching and control equipment has incorporated developments in digital technology. This stream of innovations has generated a huge increase in network capacity. Indeed, data transmission circuits were no longer exclusive for the duration of each call. Many messages could share the same circuit because each digital data packet contains information concerning the route it should follow.

In the 1980s, fibre-optics were introduced to channel long distance communications. Fibre-optics were also later used for final terminal connections. Their enormous capacity and invulnerability with respect to electromagnetic interference meant firstly a reduction of average call cost and, secondly, an improvement in the service quality. In any case, fibre-optics cable has a very high fixed installation cost (underground channelling) and, for that reason, the operators offers all sorts of supplementary services (Cable TV, Internet, interactive video, etc.) in order to increase traffic. Despite the above, given that the most extended means of transmission is still copper wire, the large telephone operators focused their research on technology capable of squeezing the very last drop out of the existing copper wire network. As a result, the ADSL (Asymmetrical Digital Subscriber Line) technology was born, the level of performance of which attempts to mimic that of fibre-optics wires.

It was also in the 1970s that data communication services came into being. The large finance firms and public administrations were the first users of this technology, whose origins date back to the 1950s. In general, they used private networks with exclusive protocols to connect computer terminals. To all events, it was soon realised that the new data transmission service could also be offered over the traditional telephone network, provided that its capacity was increased. This new market was occupied by companies from the IT sector. The so-called Integrated Services Digital Network came into being then, combining telephony and internet services.

Control systems comprise another supplementary area of technological development. Managing traffic is a major operating problem of communications networks as the output cannot be stored for long. The introduction of intelligent network systems and synchronous transfer mode switching in the 1990s reduced the operational cost of central control systems. At the same time, it improved the degree of use of installed capacity, in an environment in which the services offered have becoming increasingly diverse.

Having briefly discussed the main features of a telecommunications network, to model it from the standpoint of the fund-flow model would require the following assumptions to be made:

1. The outflow is the different types of telecommunications traffic that use the signal transmission infrastructure. Such services may be data transfer, telephone calls, videoconferencing, radio or TV broadcasting, amongst others.
2. For the sake of simplicity, no inflows are taken into account, not even power. In the case of transport, such an assumption would perhaps be unacceptable, but it is for telecommunications networks.
3. Only the fixed funds are considered. As expected, fixed funds embrace all the systems belonging to the switched and sharply defined hierarchy of telecommunications network. Depending on the model, one or more of the different segments of this complex telecommunications network may be emphasised. Nonetheless, a minimum requirement is to consider the individual line that goes from the network to the customer, in addition to the segments that carry the communication traffic between the different hierarchical nodes.

The functional of the different telecommunications services could be written as follows:

$$\left[e_1^s(t), e_2^s(t), \ldots, e_n^s(t), \ldots, e_N^s(t); S_s(t)\right]_0^T$$

with n=1,2, ..., N; s=a, b, ..., z; and $0<t<T$.

where,

1. Functions e_n^s (t) refer to the fixed funds that make up the n (n=1,..., N) subsystems (terminals, distribution equipment, switching stations, and so on) into which the network has been divided. These subsystems are required to render the different telecommunications services offered.
2. Function $S_s(t)$ denotes the outflow. It comprises the traffic dispatched by the network over a given interval of time (e.g. one year). This traffic consists of the different kinds of s (s=a, ..., z) communications services.
3. The term $0<t<T$ details the length of time the communications processes last.

For the sake of simplicity, the human work fund (employees engaged in control and maintenance activities) and inflows (basically electrical power) have been eliminated from the model.

The elementary process comprises the sequence of fixed funds required to transfer a unit of information of a given type of traffic. As known, this sequence may be understood as a set of intermediate stages, each associated with specific fixed funds. In effect, considering the simple case of an ordinary telephone call (the elementary process) that lasts on average T=200 seconds, this call may be broken down into the first 10 seconds to establish the link between the caller and the receiver, that is, to open the communication channel through a switching station. The remaining 190 seconds are invested in the conversation between the two interlocutors. Thus the telephone call considered comprises two consecutive stages: ε_1=10 and ε_2=190 seconds. Four funds are involved in the call: two terminals, a switching station and the wire that joins the interlocutors. The activity times of these funds are shown in figure 60.

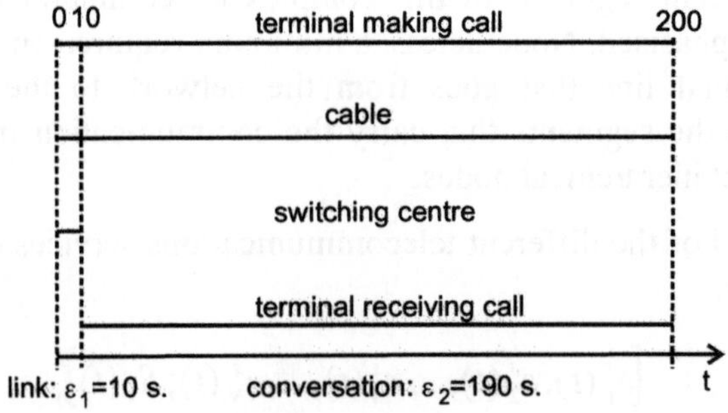

Fig. 60. An elementary process: a telephone call

The calling terminal and the line connecting it to the receiver are engaged throughout the whole course of the elementary process. To simplify matters, let it be assumed that the switching centre immediately receives the signal and requires 10 seconds to open the line between caller and receiver. As stated above, the average conversation lasts 190 seconds. At the end of the telephone call, all the funds are left idle (until the next call is made). Such an interval of inactivity is particularly costly in the case of the switching centre because it is mostly idle during the call. Telecommunications networks could then be said to suffer from idle times and under-use of funds. To tackle this issue, let us examined the following concepts and notations:

1. Vector $\varepsilon = [\varepsilon_1, \varepsilon_2, ..., \varepsilon_n]$ indicates the length of the stages comprised in the elementary process. These stages occur consecutively. Clearly, $T = \varepsilon_1 + \varepsilon_2 + ... + \varepsilon_n$, where T is the full duration of the elementary process. Each ε_i element represents the time required to complete the operations of a given stage i.
2. Terms ε_{ij} show the time the fund j (j=1, ..., J) is active in stage i (i=1, ..., I) of the elementary process.
3. Given the above elements, the expression $t_j = \sum_{j \in E} \varepsilon_{ij}$ denotes the absorption (or activity) time of fund j in the elementary process. As expected, E represents all of the stages of a telephone call.

The rate of telephone call traffic, measured in terms of the number of calls per hour (the unit of time considered herein), may be defined as:

$$A = \frac{C}{T} = \sum_{j=1}^{j=J} t_j$$

C being the number of calls, T=1 and t_j the periods of absorption of the j funds involved. In other words, the intensity of traffic is equivalent to the sum of the intervals, in an hour, when the funds are active. Consequently, the traffic is equivalent to the product of the number of calls and their average duration. Nonetheless, not all calls run immediately and so the number of these dispatched will be lower. This means that a certain volume of traffic is lost.[7]

Besides, ς_j which represents the specific scale of the j type funds should be added to the above formula. That is, the minimum usable fraction of a j-fund: a channel of a transmission line, a gate in a switching station, a portion of cable bandwidth, and so on. This fraction of a given fund may also be called an *Elementary Fund.* For the sake of simplicity, it is assumed that each elementary fund can only simultaneously handle one elementary process. For example, if a coaxial cable has a capacity of ς_j=7680 channels it is capable of carrying this number of telephone calls at the same time.

Taking the data given above, the switching centre would only be active for 10 of the 200 seconds of the whole elementary process. To ensure that its capacity is fully taken up, 20 more lines would need to be connected to

---

[7] If it is assumed that the call traffic follows a Poisson type distribution, the number of funds available and the volume of traffic supplied is taken into account, the so called Erlang Law could be obtained which indicates the rate of calls lost by a telephone network.

the centre. Or, put in another way, the facilities considered would permit 18 calls an hour to be made (18·200=3600 seconds/hour), even though the switching centre would only be operational for a total of 180 seconds. Thus 3600/180=20 is the number of lines the station should manage so as not to remain idle at any time. If the switching centre is simultaneously capable of managing, for example, 50 thousand different calls, then by multiplying this number by 20, the maximum number of lines the centre may handle is obtained (one million). This would correspond to 18 million calls per hour.

As can be seen, the reduction in idle time and the enlarged capacity of the switching centre and other fixed funds (cable) trigger a rise in the available traffic volume or capacity with virtually no increase in the number of funds. The economies of scale are staggering. Notwithstanding, it must be pointed out that these calculations are too simplistic. They have been made under the assumption that the telephone call traffic is regular and sustained. A more realistic approach should therefore be based on a specific demand profile over time and the number and capacity of the funds involved. Moreover, this approach must be based on specific conventions regarding acceptable waiting times. Meanwhile, in terms of the model presented herein, the *Saturation Scale* (S_S) for the funds involved may be defined by the following expression:

$$S_S = \min\left[{}^{1}\!/_{t_1}, {}^{1}\!/_{t_2}, \ldots, {}^{1}\!/_{t_J}\right] \cdot T = \min \alpha T\left[{}^{1}\!/_{t_1^*}, {}^{1}\!/_{t_2^*}, \ldots, {}^{1}\!/_{t_J^*}\right]$$

T is one hour, i.e., the same unit of measurement as used for the absorption times. As can be seen, in these formulae the inverse of the respective fund absorption times (t_1, t_2, …, t_J) have been broken down into two terms: the common parameter α and factors $1/t^*_1$, $1/t^*_2$,…,$1/t^*_J$. The product of αT and each $1/t^*_j$ indicates the maximum number of calls that can be handled by a given fund (in one hour). The lowest of these numbers (18 in the proposed example) establishes the saturation scale for the existing system, even though some of the funds may still have surplus capacity. This is the case of the switching centre which is capable of handling up to 360 calls per hour. In general, this scale defines a technical limitation to the system. In the current example, there is inefficiency. It suffices to replicate the S_S value until saturating the highest capacity fixed fund to achieve an efficient scale (Pannone 2001: 460).[8]

8 There is no doubt that a reduction of switching time, or the multiplication of the number of channels of certain funds, or a change in the average length of call, etc., will alter all of the above results.

If the network can carry different types of traffic, that is, voice services, data transmission or an exchange of multimedia products, involving funds with their own individual absorption times, a set of saturation scales will be obtained. There will be a particular scale for each of the services offered. Within this context, the operator's aim of maximising income could be solved by way of a complex lineal programming problem. This procedure seeks to adjust the scales of the different kinds of services to the demand structure, while their prices are borne in mind along with the provision and features of the fixed funds. As expected, these latter act as constraints (Marini and Pannone 1998: 186-9; Pannone 2001: 460-1).

6 Costs

Cost is the value of the resources required to perform an economic operation. Economic analysis has been traditionally concerned with the way costs react to changes in the level of production. Now the analysis is somewhat different: the aim is to ascertain the behaviour of costs over time, to be more precise, the behaviour of costs according to the pattern of the deployment of elementary processes. The definition and content of costs are shown in table 5.

Table 5. Costs: definition and content

		Cost element		
		Flows	Human work fund	Fixed funds
Level of cost definition	Company or group	Energy, spare parts, raw materials, etc.	Top management and administration	General equipment, resources and facilities
	Production plant or establishment		Management, administration and production plant control	Specific plant equipment and facilities
	Production line (elementary process)		Shop-floor workers and supervisors	Specific production line equipment

On looking at the table, three different levels and types of cost are distinguished. As is obvious, if the establishment houses one single production line, the costs of the plant and production process are the same. In such a case, the cost of the human work fund may be split into several large items (e.g. shop floor workers, supervisors and administration) because the tasks of these employees are very distinct. With respect to the expenditure generated by fixed funds, there are conventional rules to work out and distribute them amongst the units of outflows.

One way of approaching costs from a funds and flows outlook would be by taking the $\boldsymbol{P}_{j+k,s}$ table. As seen in the first chapter, this table indicates the amounts of inflows and service times of funds per phase of the elementary process. If these quantities were multiplied by their unitary prices, real or estimated, the table of costs per elementary process (or outflow unit) would be obtained. Another way was proposed by Zamagni (1993: 349-350). In this case, a function is built which displays, for each instant of the duration of the process, the amount paid for the inflows consumed and the fund services provided. On assuming that such payments are made at the very moment at which the elements enter into the process, the function cost takes the following form:

$$\Omega(t) = \sum_{k=1}^{k=K} p_k \cdot F_k(t) + w \cdot L(t) + \sum_{j=1}^{j=J} \xi_j \cdot S_j(t)$$

In the above expression, p_k denotes the price of the kth input flow ($k = 1, 2, ..., K$), w_j the wage rate of the human work fund and ξ_j the price of the services of the jth fund ($j = 1, 2, ..., J$). A variant on this type of cost analysis considers that the money paid for the purchase of flows or fund services may be paid at a different moment to when this element participates in the process. To such ends, all of the figures must be adequately updated. For the sake of simplicity, it is assumed that the rate of interest (r) will be constant over time. Thus, the following expression with continuous time has been proposed (Tani 1988: 12):

$$\Omega(t) = \sum_k p_k \int_0^T e^{r(T-t)} dF_k(t) + \sum_j \xi_j \int_0^T e^{r(T-t)} U_j(t) dt$$

Nevertheless, an analysis of these cost expressions would be rapidly exhausted. It is more interesting to study the repercussions on average total cost of the different forms of deploying an elementary process, i.e., sequential, parallel or line. Models developed later are therefore characterised by the crucial role given to the time dimension.

6.1 Cost and the deployment of the process

Sequential deployment is the strictly unitary consecutive activation of the elementary processes. Each task and/or phase is then performed in full by the same worker, prior to going on to the next. An analysis of the costs associated with a sequential process is based on a simplified model with the following assumptions (Piacentini 1989: 164-171, 1995: 473-476):

1. The elementary process generates a single output. This assumption is maintained over the following pages.

2. H indicates the duration of annual production activity, measured in hours.
3. The annual payments for the services of the human work funds are called W.[1] Consequently, the total labour cost per output unit is $w \cdot T$, w being the total amount paid per hour in salaries: $w=W/H$.
4. There are diverse tools and machinery.[2] These fixed funds are used in one or more of the process tasks.[3] The vector $[\sigma_1, \sigma_2, ..., \sigma_J]$ represents the annual cost allocation for the use of the j funds ($j=1,2, ..., J$). The cost of the jth fixed fund is therefore equal to,

$$\sigma_j = A_j \cdot T / H_j$$

A_j being the charge for depreciation[4] and H_j the number of hours this given fund is used per annum ($H^T \geq H_j$, where H^T=8,760 hours).
5. Some inflows are required to produce each output unit. The cost of inflows is given by,

$$\sum_1^K f_k \cdot p_k \quad (k=1, 2, ..., K)$$

in other words, the technical coefficients of the flows entering the process (f_k) multiplied by their respective exogenous prices (p_k).
6. For the sake of simplicity, it is assumed that the product is sold immediately after it is finished. Moreover, lapses of time between the

1 This remuneration includes the wages of all of the workers involved in the process. The model is simplified by not differentiating the different levels of remuneration.

2 The assumption aims to make it clear that it is not a highly mechanised process. Basically there are small steam engines, electrical motors, and so on, giving the driving force needed to power the mechanical devices which perform simple productive operations such as cutting, drilling or sawing, as well as to power mills, furnaces, etc.

3 By assumption there is no multiple capacity funds. See chapter 3.4.

4 The term A_j is equivalent to

$$\frac{(1+r)^{n_j} \cdot r \cdot M_j}{(1+r)^{n_j} - 1}$$

where M_j is the initial value of the tool or machine, n_j the number of years of useful life and r the interest rate. It should be pointed out that, for the sake of simplicity, the following assumptions have been made: the residual value is taken as zero; the effect of the varying use intensities is ignored; and maintenance and repair flows have been also included.

purchasing of flows and payments for depreciation and wages, and the instants these elements intervene in the process are not considered.

The *Average Total Cost* (henceforth ATC) of a sequential process will be then given by,

$$ATC = \frac{\sum_{1}^{J} \sigma_j + W + H \cdot \frac{1}{T} \sum_{k} f_k \cdot p_k}{H \cdot \frac{1}{T}} = T \cdot \frac{1}{H} \cdot \left(\sum_{1}^{J} \sigma_j + W \right) + \sum_{k} f_k \cdot p_k$$

In this equation, the output produced (Q) is expressed as the outflow per unit of time multiplied by the duration of the period of production (H). That is, $Q = H \cdot 1/T$. This expression shows the influence of the time dimension on average total costs.

Human work is the most significant cost item if the sequential process runs in a workshop with a master and some apprentices provided with a certain degree of mechanisation and a centralised source of energy, and the output being non-standardised products manufactured either individually or in very small batches. In that case, the above expression of the ATC may be rewritten as follows,

$$ATC = \frac{W + H \frac{1}{T} \sum_{k} f_k p_k}{H \frac{1}{T}} = T \frac{1}{H} W + \sum_{k} f_k p_k$$

In this version of ATC, the term referring to fixed funds has been eliminated. The remaining expression only includes expenditures to hire human work and buys inflows.

As known, in a craft method of production tasks are mastered after a long period of apprenticeship. Every craft-man is actually able to execute a wide range of tasks and very often, there are workers who stand out for their exclusive knowledge and experience. In this sense, it must be remembered that (re)creating a team of craft-men is a formidable task. The singular way such an establishment adjusts demand changes is then not surprising. When demand increases, the workers are urged to speed up the rate of work and/or extend the working day (J). With respect to the first option, it should be realised that the duration of a craftwork process is not a merely technical issue: T mostly depends on the degree of diligence with which the tasks are done. With regard to the latter, it should be pointed out that a longer working day will be easily accepted because of close personal relationships between workers in a workshop. To sum up, an increase in daily production (J/T) may be achieved by the reduction (or lengthening) of the duration of the process (or the working day).

On the other hand, a reduction in demand pressures the workshop to work slower and/or fewer hours. In any case, adjusting to market variations will not lead to significant changes in the number of workers employed.[5] As skills are acquired through long years of experience, shedding staff is not advisable. By the same token, it is hard to enlarge the number of workers: it is feared that the demand situation will probably return to its lower level. It is the senseless to have another worker, probably not still adequately trained. Therefore, the supply of human work funds units will slowly increase.

The top of figure 61 represents the relationship amongst the level of production per unit of time and the duration of the elementary process. The relationship between the latter and average total cost is shown at the bottom.

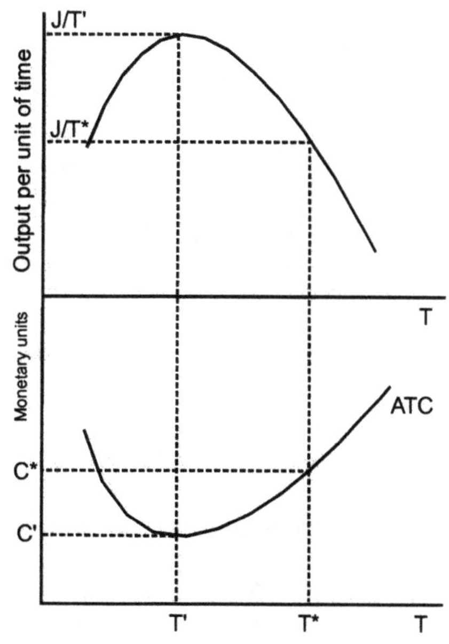

Fig. 61. Costs and the pace of production

In figure, the value T^* indicates the normal duration of the production process, i.e., the time it takes to obtain one output unit if craft-men work at the usual rate. In this case, J/T^* and C^* respectively denote the levels of

5 The employment of a new apprentice (the kind of decision normally delayed as long as possible) would not affect the essence of the argument.

production and cost. However, it is possible to accelerate the activity to a certain extent. On doing so, the position of T^* would move towards the origin of the *x*-axis until reaching T'. This point corresponds to the minimum level of average total cost (C') and the maximum level of production per unit of time (J/T'). It is also associated with the minimal duration of the elementary process, bearing in mind the procedures and techniques employed. Attempting to work still faster, or reducing the duration of the elementary process to below T', would be counterproductive: there would be an accumulation of errors and faults due to haste, increasing the number of defective output units. The curve therefore falls to indicate a lower number of *saleable* units. This will of course have a negative impact on the level of average total costs.

In order to draw the ATC curve as T^* falls the following assumptions are established:

1. For the sake of simplicity, it is assumed that the cost of flows remains unchanged.
2. φ denotes the relationship between the good enough smallest and normal duration of the elementary process: $\varphi = T'/T^*$ and, given that $T' < T^*$, then $\varphi > 0$.
3. A simple way to represent the quality lost of the product is to set up a parameter Ψ indicating the accumulation of defective output units which could not be sold. The cost will be progressively higher, the more it is attempted to reduce T^*. Therefore, the surcharge due to flawed units of output is $\Psi/\varphi T^*$.

The behaviour of average total costs prior to modifying T is given by the expression:

$$ATC = wT + \sum_k f_k p_k$$

In placing the two terms defined above, it becomes:

$$ATC = \varphi w T^* + \sum_k f_k p_k + \frac{\Psi}{\varphi T^*}, \ \varphi > 0$$

This equation expresses the relationship between average total costs and the variable φ, which regulates the length of the elementary process. Rearranging the terms,

$$ATC = \frac{1}{\varphi}\left(\varphi^2 w T^* + \frac{\Psi}{T^*}\right) + \sum_k f_k p_k, \ \varphi > 0$$

Denoting the constant terms with $a=wT^*$, $b=\Psi/T^*$ and $c=\sum_k f_k p_k$, the final expression is,

$$ATC = \frac{1}{\varphi}(a\varphi^2 + b) + c\,,\ \varphi > 0 \qquad [1]$$

which corresponds to an equation which the following generic form:

$$y = ax + \frac{b}{x}$$

For positive values of φ, the expression [1] presents a *U*-shaped pathway.[6] Or, put in substantive terms, the reaction of average total costs to a lessening of the elementary process is favourable until reaching the minimum point associated with the notional maximum pace of production. But any attempt to go further in the reduction process will give rise to a large number of defective units. This will result in an increase in costs.

The curve relating ATC and the duration of the elementary process will move upwards (downwards) according to greater (lower) salaries, prices of the inflows or quantities required of these per output unit.

In a competitive market, several workshops with sequentially organised production processes should face flexible prices (Nell 1993). It has been noted that this price flexibility behaves as following: the more acute the *rate* of growth (fall) in demand ($\hat{d}$) with respect to prior period, the more intense the response of prices ($\hat{p}$) in the same direction. See figure 62 (adapted from Sylos Labini 1980: 185). Given that by definition, $\hat{p} = f(\hat{d} - \hat{s})$, $f' > 0$, the greater the increase in demand, the worse the supply rate of change ($\hat{s}$) can adjust. This fosters price increases. On the other hand, a fall in demand tends to lower prices, despite the resistance from producer organisations. In both cases, prices overreact.

All that has been said until here reflects the well-known idea that small changes in demand can be managed by changing the pace of work and/or the length of the working day. This adjustment may be fast and, consequently, prices are unaffected.

---

6 This type of function is decreasing (rising) in the left (right) side of $x = \sqrt{b/a}$. So, at this abscissa point, these functions have a minimum.

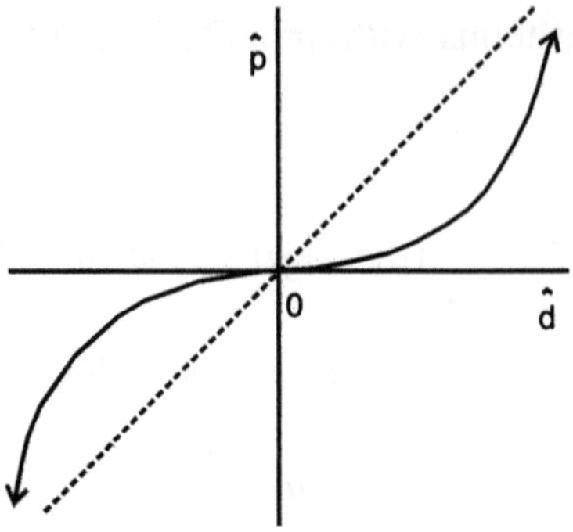

Fig. 62. Price-demand reaction

Now let us come back to the line process. As has already been stated in previous chapters, Classical school economists were greatly impressed by the levels of productivity achieved by the Factory System. This new industrial production system combined the advantages of the technical division of work with those of the line deployment. Workers improved skills and dexterity and became more productive. Idle times for workers and machines alike suffered from drastic reductions. It also opened the doors to the extensive automation of production processes.

The line deployment implies a greater output per unit of time and lower average costs with respect to the sequential process, as Babbage observed in his day. An important reason for that is the existence of idle time in the latter. Despite the fact that this is an evident outcome, it could be again illustrated using the expressions of costs. As it has been explained, each task (d_j) may be expressed as the product of two: $d_j = \delta^* \cdot v_j$. The former (δ^*) indicates the maxim common divisor of the duration of the different tasks. The latter (v_j) shows the number of fund units purchased or rented. Given that the output per hour of a line process is $1/\delta^*$, then $1/\delta^* \cdot H$ indicates the total annual output. This level of production is higher than that a sequential process can achieve. In effect, let

$$\sum_h i_h$$

be the sum of the intervals of idle time between the phases of the elementary process. Given that $1/T$ is less than $1/\delta^*$ because,

$$T = \sum_j v_j \cdot \delta^* + \sum_h i_h \,.$$

To verify whether or not the increase in output from a line deployment generates economies of scale the costs of the two methods of production should be compared. To all events, a line layout requires several fund units. Let us suppose that the number of fund units required is given by the vector $(\phi_1, \phi_2, ..., \phi_J)$ being,

$$\phi_j = \frac{d_j}{\delta^*} = v_j, \text{ with } j = 1, 2, ..., J$$

Taking the assumptions established in the case of sequential production, the average total cost of line deployment is given by,

$$ATC = \frac{\sum_j \phi_j \sigma_j + W \sum_j \phi_j + \frac{1}{\delta^*} H \left(\sum_k f_k p_k \right)}{\frac{1}{\delta^*} H}$$

Or, reorganising terms,

$$ATC = \delta^* \frac{\sum_j \phi_j \sigma_j + W \sum_j \phi_j}{H} + \sum_k f_k p_k$$

Comparing the two expressions of cost implies, in first place, comparing

$$T \sum_j \sigma_j$$

and

$$\delta^* \sum_j \phi_j \sigma_j$$

Thus, as seen above,

$$\delta^* \sum_j \phi_j \sigma_j = \sum_j d_j \sigma_j = T \sum_j \frac{d_j}{T} \sigma_j .$$

Given that $d_j/T<1$, we finally obtain:

$$T \sum_j \sigma_j > \delta^* \sum_j \phi_j \sigma_j .$$

Secondly, it should compare,

$$T \cdot \frac{W}{H}$$

and

$$\delta^* \cdot \sum_j \phi_j \cdot \frac{W}{H}$$

From here it is concluded that,

$$T > \delta^* \cdot \sum_j \phi_j = \delta^* \cdot \sum_j v_j .$$

As expected, the average total cost of the line process is lower than that of the sequential process. This conclusion obviously relies on the assumption that

$$\sum_j d_j < T$$

that is, there are intervals of inactivity within the elementary process which cannot be avoided if the elementary process is deployed in a sequential manner. Given the goal of economic profitability, its deployment in line will give rise to the introduction of organisational and technical improvements. As a result, the number of its manufacturing stages along with its temporal gaps will be reduced.[7]

6.2 The cost and pace of production

If the speed of the process changes, costs also change. This subject may be analysed by a simple model containing the most relevant technical and economic variables that intervene in a line process (adapted from Petrocchi and Zedde 1990; Piacentini, 1997). Following the analysis started in the preceding pages, take a manual assembly process characterised by the next assumptions:

1. The process contains one single phase divided into tasks performed by specific funds.

---

7 From a historical point of view, the occurrence of process innovation frequently ends up changing the architecture of the manufacturing process. As a consequence, the number of stages is reduced and progress is made towards a greater continuity of the production stream (Utterback 1994).

2. Each task requires one single worker with certain tools.
3. The annual total for the services of human work is denoted by W.
4. A is the annual charge for the depreciation of the fixed capital assets.
5. H represents the total number of hours of activity per annum. So $H=J\cdot D$, J being the duration of the working day and D the number of work shifts in a calendar year.[8]
6. The lapse of time between the purchase of the flows and the payments for depreciation and salaries, and instants these elements intervene in the process, are not considered.

The average total cost per product unit is the annual cost of inflows and funds services divided by the total output produced in one year. That is,

$$ATC = \frac{W + A + \frac{1}{c^*} H \sum_k f_k \cdot p_k}{\frac{1}{c^*} JD} = c * \frac{W + A}{H} + \sum_k f_k \cdot p_k$$

$\sum_k f_k p_k$ ($k=1, 2, ..., K$) being the cost of inflows. In this expression,

1. Given that $H=J\cdot D$.
2. Denoting w as the total wages per hour ($w=W/H$) and $W/m\cdot H$ being the salary per worker per hour (m the number of workers on the line).
3. α being the charge for depreciation per hour ($\alpha=A/H$).

Then, the ATC becomes,

$$ATC = c^*(\alpha + w) + \sum_k f_k \cdot p_k$$

If it is now remembered that a great level of production may be achieved by:

1. Reducing c^*, and
2. Reducing the cycle by cutting back the service time (s_i) and/or idle time of funds (e_i),

the following final expression is obtained,

[8] This should have been written as $J=\beta\cdot J^*$, $0<\beta\leq 1$, to encompass the number of effective working hours. Indeed, interruptions may occur for diverse reasons, programmed or not. It must also be added that D coincides with the number of days of labour, if these includes one single shift.

$$ATC = \frac{\lambda}{n}\left[c^{*}(\alpha + w) + \sum_{k} f_k \cdot p_k\right] \quad \text{with } n>1 \text{ and } 0<\lambda<1$$

This is the sought after cost function. It brings together the different sources of cost and the strategies to increase production.[9]

Figure 63 has been built to examine the issue in greater depth. The horizontal axis refers to the rate of production while the vertical one consists of two parts: the upper section measures cost per output unit while the lower shows the gap between consecutive elementary processes (the narrow gaps are those closer to the origin). As can be seen, changes in the parameters (W,A,H,f_k and p_k) lead to the displacement of the curve without altering its general form.

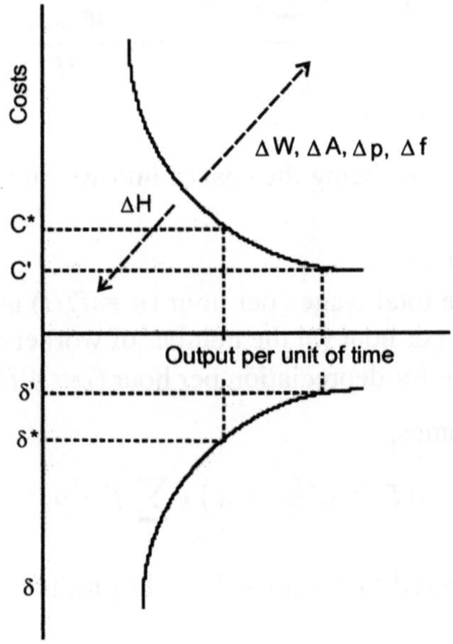

Fig. 63. Cost, gap and output per unit of time

The *normal* pace of production and cost levels are situated in a relatively narrow interval between the optimum gap δ^* (associated with C*) and the maximum pace of activity δ' (associated with C').[10]

---

9 In the expression, it has been assumed that the impacts of the increased pace of activity (the factor λ/n) do not raise the rate of depreciation of the fixed funds (α). This assumption has been established for the sake of simplicity.

It should be noted that the above cost curve is *not* the typical short-term plant cost curve. In effect, the position of the proposed curve is subordinate to the value of a large number of parameters. After fixing these parameters, the curve will then simply show the relationship between the average cost and λ/n values. Moreover, on carefully analysing the graph, the different factors affecting the cost per unit of output could be described. Some directly affect the cost, while others act through changes in the output per unit of time. All of these circumstances fall into one of the following three cases:

1. Due to time. Some institutional factors, such as the length of the working day (J) or changes to the annual working calendar (D), make the point. Others factors related to the internal organisation of the process and the capacities of the funds employed (which will alter δ^* or c^*) should be added. All of them lead to a time saving in the process. They therefore indicate the time efficiency of the process.
2. Those related to the efficiency in using materials. This is the case of the amount of flows absorbed per output unit (f_k).
3. Economic ingredients such as the prices of the funds and flows.

With regard to the process adjusted to fluctuations in demand, after having accumulated stocks of finished products, the next decisions to consider are:

1. The reduction of the monthly, weekly or daily working periods by means of temporary plant closure.
2. The slowing of the pace of activity. This decision tends to be translated into a reduction in the number of funds employed at the different workstations. In other words, some workers are laid off and machines left inactive.

If demand grows, the option is to encourage overtime, extend the number of shifts and amend the working calendar (working weekend, shortening holiday leave). Another option is to accelerate the pace of work, which could involve increasing the number of funds contracted (Bresnahan and Ramey 1994).

[10] In some figures on costs, the term δ related to the cycle has been drawn instead c to avoid confusion of this latter with cost symbols. As known, δ and c are the same.

6.3 Time efficiency and funds

The vector $[\sigma_1, \sigma_2, ..., \sigma_J]$ representing the annual cost for the use of the j funds ($j = 1,2, ..., J$) was defined in chapter 6.1. As also seen, the cost of the jth fixed fund is equal to $\sigma_j = A_j \cdot T / H_j$, A_j being the charge for depreciation and H_j the number of hours that this fund is used per annum ($H^T \geq H_j$, H^T=8,760 hours). The analysis now goes deeper taking into account the concept of (relative) efficiency in terms of time. With regard to this, we can consider two forms of *wasting* time:

1. Due to the intervals of inactivity of funds. This occurs through different working times of workstations and by other more or less transitory circumstances.
2. The duration of the elementary process is longer than in other plants. Many reasons could explain that: more intensive use of the existing funds, better organisation of the tasks, and so. The point is that other establishments, using the same basic technology, take advantage of one or more shorter tasks, which means a lower T value.

The two sources of comparative time inefficiency affect the funds, either capital equipment or human work. Therefore, given the expression for the cost of the jth fixed fund,

$$\sigma_j = A_j \cdot T / H_j$$

and denoting the A_j/H_j ratio as the cost per hour (s_j), it can be written,

$$\sigma_j = s_j \cdot T$$

The above expression quite simply indicates the cost of using the jth fixed fund by the elementary process. Indeed, it is the cost per hour multiplied by the duration of the elementary process. In order to consider the repercussions on costs of any possible lapses of inactivity, a *coefficient of use of the jth fixed fund* is defined as,

$$\varepsilon^m{}_j = t_j / T, \text{ with } 0 < \varepsilon^m{}_j \leq 1,$$

t_j being the lapse of time this given fund is active in the elementary process. Consequently, if this degree of use is combined with the above cost expression, we obtain:

$$\sigma_j = s_j \cdot T / \varepsilon^m{}_j \neq s_j \cdot T$$

This expression indicates the extra cost added by the existence of intervals of fund inactivity when comparing two or more processes.

In the case of labour, the analogous expression would be $w \cdot T/\varepsilon^l$, w being the total wage per hour and ε^l the coefficient of use of the human work fund. This expression may be accordingly modified to including the different types of workers.

With regard to the lower relative efficiency provoked by the excessive duration of the elementary process, the concept of *Elementary process saturation* can be considered (Piacentini 1987: 385; Petrocchi and Zedde 1990: 62). This term is defined as follows:

$$\iota = T^* / T$$

T^* being the duration of the elementary process referred to. It is assumed that T^* is lower than T, and then, $\iota \leq 1$.

Bearing in mind the two different forms of time inefficiency, the expression of cost of a given fund may be suitably modified as follows:

$$\frac{T^*}{\iota}\left(\frac{w}{\varepsilon^l}+\sum_1^J \frac{s_j}{\varepsilon_j^m}\right)$$

This is the *Cost of funds concerning time inefficiency*. A hypothetical representation of this singular cost function is shown in figure 64.

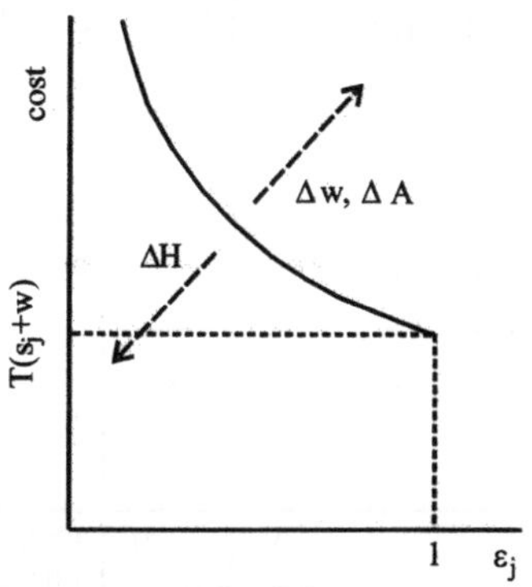

Fig. 64. Cost of funds and time efficiency

The vertical and horizontal axes represent the cost of the funds in each elementary process and the product $\varepsilon_j = \varepsilon^m{}_j \cdot \varepsilon^l \cdot \iota$, respectively. Evidently, in a line process with no loss of efficiency, the time cost for the funds would coincide with the point $\varepsilon^m{}_j = \varepsilon^l = 1$ on the x-axis associated with the ordinate point $T(s_j + w)$. Moreover, the figure shows how the curve would be displaced by changes in the parameters considered. For example, an exogenous increase in salary or in the annual charge for depreciation would mean an upward movement, while an increase in the annual number of hours of plant activity would lead to a downward movement.

The improved use of flows would usually mean a reduction in the average rate at which they enter into the process, i.e. a lower f_i value per output unit. However, it must be pointed out that the saturation factor may also influence the consumed amount of flows that can either be reduced or increased. For example, the reduction in loading and unloading times for the machinery may give rise to a fall in energy consumption, even though this higher intensity of use can lead to a greater need for repair and maintenance operations.

6.4 Short-term plant costs and price determination

In the above section, the general relationship between the level of production (per unit of time) and costs has been established. It is a short-term analysis because of it is assumed that changes in the pace of activity (c or δ) do not represent any modification to the production methods and techniques. In this context, once a certain pace of production is established as *normal* that firm may sometimes attempt to increase it. The aim is to reach the maximum level of output per unit of time as is shown in figure 65.

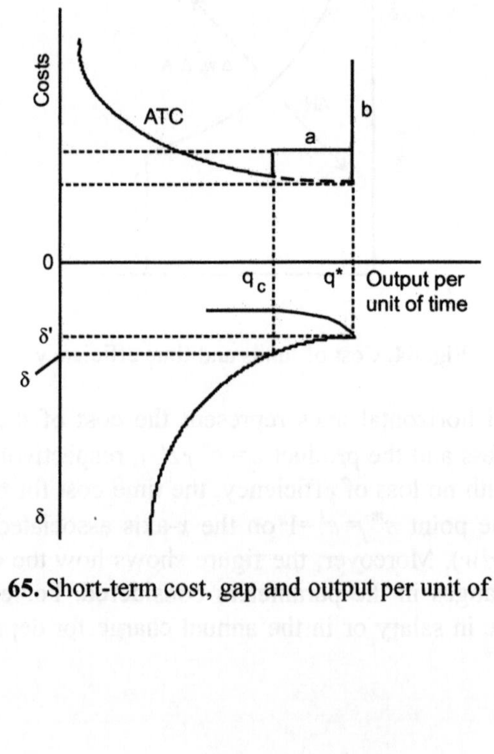

Fig. 65. Short-term cost, gap and output per unit of time

In the figure, the level of production q_c, associated with c or δ, is considered the *normal* or *standard production level* (per unit of time). With respect to the output level q^*, associated with δ', it is the maximal production (per unit of time) achievable with the facilities and technology in use. It should be noted that the normal level of activity always includes a narrow margin of idle capacity which is justified as it gives to the process a degree of flexibility.[11] For example, it allows the plant activity to cope with any small, unforeseen and transitory upswings in demand. Meanwhile, although average total costs at maximal capacity are slightly lower than normal, it must be recognised that any attempt to sustain the maximal production level permanently will end up causing bigger costs for breakdowns, errors, and so on. For this reason, section $q_c q^*$ (or segment a) of the figure is associated with average total costs slight higher than those linked to the normal level of activity. Whatever the case, it is not feasible to produce beyond the maximum technical level (point q^*). This fact is represented by a leap to unlimited growth on the average total cost curve (or segment b). As a conclusion, all points situated further than the standard production level therefore have a very anomalous character. For this, they are of little relevance for analytical purposes. Indeed, the section $q_c q^*$ will only be considered if there are good transitory prospects for demand. However, if the pressure of increased demand remains, it is possible to lengthen the working day (overtime) and/or establish a second, or even a third, shift (if the standard is an eight hour working day).

The narrowness of the interval $[\delta, \delta']$ means returns may be considered constant, or in other words, having set the normal pace of the line and, given that the maximal rate of production only occurs under extraordinary circumstances, returns may be considered as constants. Of course, changes to the capacities of the funds and/or the organisation of the process would open the door to increasing returns.

From a static point of view, the description of the relationship between the average total costs and volume of output per unit of time follows the conventional division of fixed and variable costs. Such a classification does not entirely fit with that of the funds and flows model, although a link

[11] In all the figures on costs drawn on the next pages the normal production level and the so-called full capacity level have been considered to be the same. Indeed, some authors define the normal capacity as equivalent to 70-85% of full capacity, assuming that the maximal capacity is still greater. There is no doubt that this description is fitted to the everyday life of plants. Our assumption has been made only for the sake of simplicity. Further discussion on plant cost curves may be found in Eichner (1986: chap. 3), Lavoie (1992: 118-148) and Lee (1998).

may be established between the two. Indeed, fixed costs are those generated by the fixed funds while the flows and human work fund account for the direct or variable costs. It may be then written,

$$ATC = AVC + AFC = CV/q + CF/q = \frac{w + \sum_k f_k \cdot p_k}{q_c} + \frac{\alpha}{q_c}$$

In the expression, q_c refers to the amount produced per unit of time at a given rate of activity (measured by cycle –c) while the already familiar terms w and α respectively denote the hourly wage and charge for depreciation. A standard working day is assumed (e.g. J=8 hours with one single shift and no weekend work). Because the lower q_c, the higher the burden of fixed costs, the shape of the average total cost curve is as shown in figure 66. This figure also contains the average variable cost and the marginal cost curve. Both are horizontal and overlap in their relevant segments.

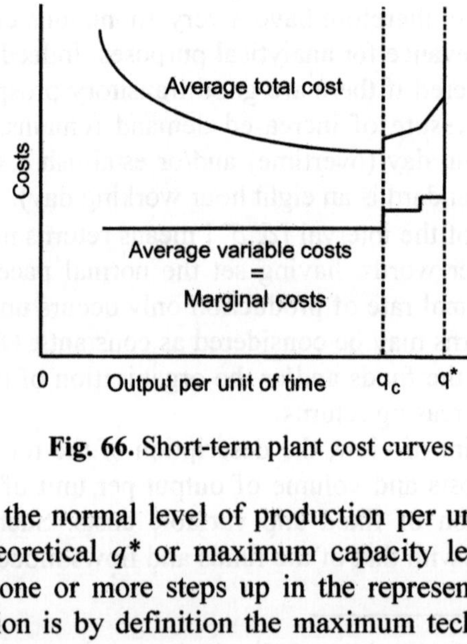

Fig. 66. Short-term plant cost curves

As known, q_c is the normal level of production per unit of time which is lower than a theoretical q^* or maximum capacity level. Beyond the q_c point, there are one or more steps up in the represented curves. The q^* level of production is by definition the maximum technically achievable. This absolute limit is represented by a vertical section of the average total, variable and marginal cost curves.

Looking at the short term, the technical coefficients could be considered as fixed. A plant is designed with a view to a given level of activity, i.e., the so-called normal level of production per unit of time. This production level also accounts for a foreseeable degree of stoppage, for both expected breakdowns and routine maintenance and repair.

Babbage's *Multiple's Principle* was introduced in chapter 2. For this author, the process of production may only be replicated in parallel, after having determined the output level per unit of time that, with a given production technique, will guarantee the elimination of all fund idle time. To avoid significant repercussions in terms of costs, the new scale will be a multiple of the former. This condition is also applied to the present analysis of the short-term plant cost curves: once the minimum of the average total cost curve has been achieved, the output per unit of time may only be replicated by a proportion expressed in integer numbers. The entire production plant will accordingly be repeated.

With regard to prices, these are determined by adding a margin to the costs. There are three main variations to such cost-plus pricing methods:

1. Mark-up pricing consists of adding a gross margin on the direct or variable costs. This margin includes the charge for depreciation. This simple principle is still used by small and medium-sized firms because its application only needs limited accounting information. That is:

$$p=(1+\gamma)\cdot AVC=(1+\gamma)\frac{w+\sum_k f_k \cdot p_k}{q_c},$$

   Direct costs include the wages of the plant workers and the cost of all inflows. The relationship between the profit margin and price is very simple:

$$m=\frac{\gamma \cdot AVC}{p}=\frac{\gamma \cdot AVC}{(1+\gamma)AVC}=\frac{\gamma}{1+\gamma}$$

   The portion of gross gains was called *monopoly grade* by Kalecki. The mark-up (γ) and margin (m) may be also related by $\gamma = {}^{m}\!/_{1-m}$.

2. Full-cost, or normal pricing, is the most realistic and widely-used method. It consists of adding a net margin to the average total cost, that is, $p=(1+\rho)\cdot ATC$:

$$p=(1+\rho)\cdot ATC=(1+\rho)\left(\frac{w+\sum_k f_k \cdot p_k}{q_c}+\frac{\alpha}{q_c}\right)$$

   It must be remembered that the normal average total costs refer to the standard rate of capacity used (q_c) and incorporate all manufacturing costs, including depreciation and general administrative overheads. This price setting method is currently used by large companies. Its practice has increasingly extended as innovative accounting techniques have permitted the general overheads to be allocated among the different goods produced.

3. The target-return pricing method consists of establishing the level of net margin according to the previously fixed target rate of return on corporate capital. This method calls for the use of sophisticated accounting tools, given that it requires working out the value of the capital employed by the company. On the one hand, r is the target rate of profit and $p{\cdot}K$ the value of the stock of capital of the firm. The profits pursued for the current period are $r{\cdot}p{\cdot}K$. On the other hand, let the capital/output ratio be denoted by $v{=}K/q_c$, q_c being the normal use of capacity (by assumption, equal to the full capacity). The target of profit per output unit to be attained will be:

$$\frac{r \cdot p \cdot K}{q_c} = r \cdot p \cdot v$$

The amount of profits per output unit is given by,

$$\rho{\cdot}ATC$$

In order to calculate ρ the last two expressions should be equated:

$$r \cdot v \cdot p = \rho \cdot ATC \qquad [1]$$

Since the normal cost price equation, $p{=}(1{+}\rho){\cdot}ATC$, may be rewritten as:

$$ATC = p/(1{+}\rho),$$

replacing the ATC term in [1], we finally obtain:

$$\rho = \frac{r \cdot v}{1 - r \cdot v} \qquad [2]$$

Hence, the relation between the full-cost and the target-return pricing is given by:

$$p = \left(1 + \frac{r \cdot v}{1 - r \cdot v}\right) ATC \qquad [3]$$

Or, in its full form:

$$p = \left(1 + \frac{r \cdot v}{1 - r \cdot v}\right)\left(\frac{w + \sum_k f_k \cdot p_k}{q'} + \frac{\alpha}{q'}\right)$$

From this expression, it can be realized that a high r value or a high capital/capacity ratio, will lead to an increase in ρ and, given the costs, also in the price. Indeed, deriving [2] respect to r:

$$\frac{d\rho}{dr} = \frac{v(1 - rv) - (-v^2 r)}{(1 - rv)^2} = \frac{v}{(1 - rv)^2} > 0$$

As this derivative is positive, ρ responds in the same way as do r, ceteris paribus. It should be added that this price determination method could be connected to the Sraffian model of general or sectoral interdependence (see Lavoie 2004: 47-8): the rate of target yield is closer to the rate of normal profit of this multi-sectoral model. In effect, the price expression [3]:

$$p = \left(1 + \frac{r \cdot v}{1 - r \cdot v}\right) ATC$$

may be written otherwise if it is supposed that ATC represents the cost of the labour fund per output unit: $w \cdot c$. Then,

$$p = \frac{1}{1 - r \cdot v} wc$$

The above expression may also be written as:

$$p = wc(1 - r \cdot v)^{-1} \qquad [4]$$

which is closer to production prices as defined by the Sraffian model of general interdependence. Indeed, as known, these latter prices are defined,

$$\boldsymbol{p} = w\,\boldsymbol{n} + r\,\boldsymbol{M}\,\boldsymbol{p}$$

$\boldsymbol{p}$ being the vector of prices, $\boldsymbol{n}$ the vector of the technical coefficients of labour and $\boldsymbol{M}$ the matrix of the technical coefficients of the means of production. This expression may be written as follows:

$$\boldsymbol{p} = w\,\boldsymbol{n}\,[I - r\,\boldsymbol{M}]^{-1}$$

which is virtually identical to equation [4].

Given that, in the short-term, prices are decided by looking at one's own costs and the prices set (and the sales conditions promoted) by rivals who offer products of a similar quality, these prices will not necessarily clear the market. Firms strive to obtain information on their competitor's costs and investment plans, at the same time using accounting tools to control the evolution of their own costs. In any case, only the leading firm may set the market prices with a certain degree of discretion. These prices are tacitly accepted by the sector. These prices guarantee a similar level of profits for firms whose level of efficiency is equal to or higher than the average for the sector.

The cost-plus approach emphasises the fact that the profit margin will be fairly insensitive to fluctuations in demand. Any adjustment is mostly quantitative (higher or lower level of production per unit of time) than affecting prices, because the goal is to maintain the margin. Otherwise, it must be remembered that many firms update prices as average total costs increase.

6.4.1 Cost and overtime

The level of plant production, per unit of time, may be expressed as a combination of the variables shown in table 6.

Table 6. The level of production of an establishment

Output per production line	=	(Hours of standard working day + hours of overtime) x Number of shifts x Production per hour
x		
Number of lines		
=		
Establishment or plant production		

There is an endless variety of economic circumstances (behaviour of demand, input costs, etc.) and institutional restrictions (legal rules, trade union agreements, etc.) which influence the activity of any plant. However, in this section the focus will be put on the pattern of working times. In this respect, taking a sufficiently long period of time (a month, a year) the observed fluctuations in the volume of output may be partially explained by changes in the working time. Indeed, overtime and the increasing of shifts are the procedures normally employed to increase production levels (Winston 1974; Betancourt and Clague 1981; Bresnaham and Ramey 1994). Evidently, these ways of lengthening the period of work have an impact on the short-term cost curves of a plant. In order to describe this, two simple models are proposed on the next pages. Both are based on the already familiar expression for average total cost, namely,

$$ATC = \frac{W + A + \frac{1}{c^*} H \sum_k f_k p_k}{\frac{1}{c^*} JD}$$

In this expression, the numerator indicates the total annual cost for the use of the funds and consumption of flows, while the denominator expresses the total annual production as the output per hour ($1/c^*$) multiplied by the length of the working day (J) and the number of shifts worked per year (D). Both J and shifts are assumed to last 8 hours.

It is convenient that the expression of ATC be written as follows,

$$ATC = c^* (\alpha + w) + \sum_k f_k p_k$$

This simplified version incorporates the number of hours of activity per year ($H=J \cdot D$) and the terms w denoting the total payroll ($w = W/H$) and α the charges for depreciation ($\alpha = A/H$), both stated per hour.

By an appropriate manipulation of the preceding ATC equation, the analysis starts with the overtime case. First of all, it may consider the average salary per hour for the whole of the working day (ω), i.e., the standard eight hours plus overtime. This may be defined as,

$$\omega = \frac{w[1 + \theta(1 + \xi)]}{1 + \theta}, \ \theta > 0 \text{ and } \xi \geq 0$$

θ being the lengthening of the working day, $\bar{J} = (1 + \theta)J$, and ξ the extra wages paid for overtime. This remuneration is assumed to be equal to, or higher than, the current salary. However, his assumption may not hold true under different circumstances.

The second step is to compare the average total costs with and without overtime. The difference (π) between the two is given by the expression:

$$\pi = c^* \left[\frac{w[1 + \theta(1 + \xi)]}{1 + \theta} + \alpha \right] - c^* (w + \alpha)$$

It must be pointed out that the flows are not included in this expression. It is therefore assumed that the rate of consumption of the flows is not affected by the lengthening of the working day. At the same time, the depreciation for funds is supposed to be identical for both the standard working and overtime hours. Nonetheless, if the outlay for maintenance and repair were to increase it could easily be included in the model.

The final result is,

$$\pi = \frac{\theta\xi}{1+\theta} wc *$$

The short-term global plant cost (or cost curve including overtime) presents the following form:

$$ATC = \left(1 + \frac{\theta\xi}{1+\theta}\right) w + \alpha$$

The factor,

$$\frac{\theta\xi}{1+\theta}$$

quite plausibly has a value close to zero. This will raise the average total cost curve to above the ATC curve without overtime. With respect to the average variable cost curve, which coincides with that of marginal cost, it has a constant value,

$$AVC = MC = \left(1 + \frac{\theta\xi}{1+\theta}\right) w$$

All of these curves are represented by the unbroken lines in figure 67.

One final consideration: firms take advantage of overtime since it is relatively quick to implement, can be performed by the ordinary company workforce and requires no changes to capital equipment.

6.4.2 Cost and shifts

The second case to be investigated is referred to shifts (each, presumably, of 8 hours and, then, a maximum of 3 per day). The analysis of short-term plant cost curves of increasing the number of shifts is somewhat more complex. To start with, expanding the number of shifts (D) will probably imply paying the workers at a higher rate for the personal annoyances of the second and, especially, the third shift, along with those on weekends and holidays (Betancourt and Clague 1981: 221-231). These additional shifts will also require higher amounts of certain flows per output unit. An example of this is the big expense on electricity on the night shift because the work area has to be adequately lit. Despite the fact that this represents an extra cost, it is also probable that power will be supplied at a cheaper rate per unit at night.

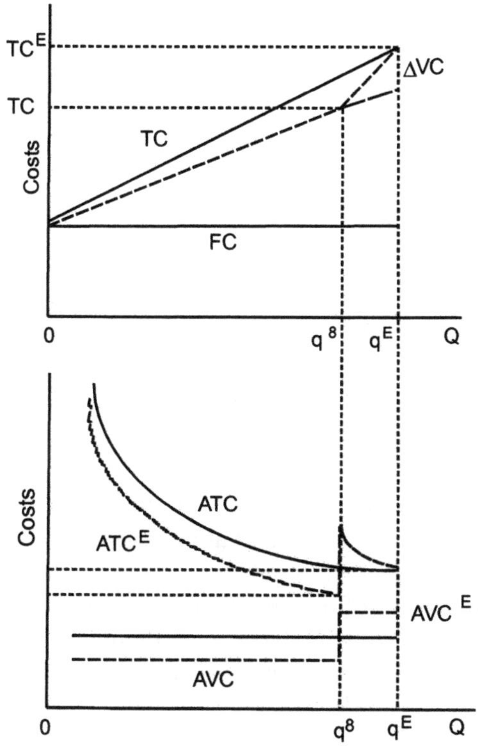

Fig. 67. Overtime cost curves

In addition, extending the number of shifts alters the hourly charge for capital depreciation. At this point, a distinction between depreciation for functional wear and tear and for obsolescence should be made. The former is measured in terms of calendar time according to the expected useful life of a given fixed fund. The number of shifts will not excessively affect this temporal rate. Rather, the forecast of obsolescence is independent of the temporal span of activity of the fixed fund. Then the greater the number of hours used by the fixed fund, the lower the hourly charge for depreciation, before the unit is withdrawn for obsolescence.

To finish with, it is plausible that extra shifts be carried out by lower skilled workers, because of the extemporaneous nature of such working hours.

To find the short-term cost curves, the costs of the first and second shifts should be compared.[12] The point is: the option for setting up a second shift will be chosen if $ATC^2 < ATC^1$ (where the superscripted number indicates the number of shifts actives), or in other words, $ATC^1 + \pi = ATC^2$. Having considering the appropriate expressions of costs:

$$\pi = \frac{c^*}{2}\left[w(2+\varepsilon) + (1-\zeta)\alpha\right] + (1+\vartheta)\sum_k p_k f_k - \left[c^*(w+\alpha) + \sum_k p_k f_k\right]$$

is finally obtained in which the considered ATC^2 has the following features:

1. The term $\varepsilon \geq 0$ represents the higher wages paid to the workers on the second shift. That is, $w+(1+\varepsilon)w$, w being the current wage rate. The global remuneration will therefore be equal to $w(2+\varepsilon)$.
2. The element $\vartheta > 0$ indicates the higher expenditure for inflows per output unit. This amount is not broken down into individual flows because only the upper outlay is important.
3. The factor $0<\zeta<1$ shows the fall in the hourly charge for depreciation (α), due to the fact that more hours are worked.
4. The production is twice that of one single shift: $1/c^* \cdot J \cdot 2D$. For this reason, the expression of average total cost appears divided by ½. This assumption simplifies the analysis, even though it is likely that the total production at the end of the working day will be slightly less.

The final expression is,

$$\pi = \frac{c^*}{2}\left[\varepsilon w - \alpha(\zeta - 1)\right] + \vartheta \sum_k p_k f_k$$

With respect to the value of the parameters of the above expression, it may be assumed that the extra salary (ε) paid will be more or less offset by the fall in the charge for depreciation (α). At the same time, the presumably higher rate of consumption of flows into the second shift (assuming that their cost does not vary) will tend to raise the average total cost.

Graphically, the short-term plant cost curves with two shifts would appear as shown in figure 68. This figure has been designed on the basis of the following assumptions:

1. The volume of production per unit of time is the same for the two shifts.
2. The consumption of one or more flows is higher, thus increasing the average variable cost. This is shown with the change of gradient of the

[12] The analysis could be extended without problems to the third shift case.

total cost curve and the layout of the corresponding sections of the average variable cost curves for the different shifts.

3. The average total cost lessens in changing from one to two shifts. Indeed, fixed costs are divided by a greater number of output units. The average total cost curve for the two shifts therefore envelopes the curve for one single shift.

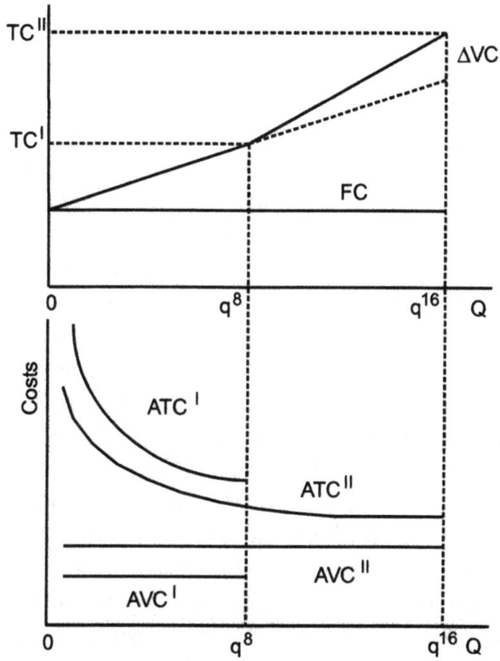

Fig. 68. Total and average total costs with two shifts

6.5 Economies of scale and indivisible funds

Generically, two *vectors* $\boldsymbol{X}$ and $\boldsymbol{Y}$ will have a scaled relationship if a link between them as $\boldsymbol{Y}=\varepsilon\boldsymbol{X}$, with $\varepsilon>0$, may be established. This very simple nexus may occur with a limitless number of variables associated with all sorts of phenomena. In the case of microeconomic production theory, the scale relationship that has raised the greatest interest is that established between the levels of production and costs in comparing various plants. This subject is known as *economies of scale*: the higher the output volume per unit of time the lower the average total cost. In general, these economies are interpreted as the consequence of greater efficiency of huge production levels. To all events, economies of scale are conceived as an empirical phenomenon. This fact was observed back in the very earliest days of the science of political economics, although the concept has undergone subtle change over the years.

For the Classical school authors, a high output volume per unit of time was an essential, though not sufficient, condition for greater efficiency (Scazzieri 1981: 93). The most refined expression of this idea was Babbage's principle of the multiple: despite this being an indisputable fact, there is no simple relationship between scale and efficiency of production, because of it moves in leaps which are caused by the condition of replicating only balanced processes in order to maintain the level of efficiency. Furthermore, these economists foresaw possible difficulties in fully benefiting from the advantages of big scale: markets too narrow, problems with the management of such firms and so on).[13]

However, over the decades, the concept of large scale of production has changed its definition: the greater scale has been seen as the *cause* of the (assumed) greater efficiency of major establishments. Massive production has thus become synonymous with increasing efficiency. Therefore, it is no surprise that economies of scale are represented with smooth and uninterrupted cost functions. These cost functions at first fall and later rise (diseconomies of scale) according to the theory of changing yields of perfectly divisible processes. At the same time, this hypothesis has been established assuming that a change in the scale of production is feasible without affecting the proportions of the factors of production, or in other words, without changing the manufacturing technique.[14] In any case, conception of economies of scale is nowadays surrounded by several ad hoc assumptions. This leads to the concept of *optimum* scale. Nonetheless, in empirical terms it would be more appropriate to talk of *critical* scale.

Leaving aside the theoretical discussion, the lower average total cost coupled to a greater level of plant production is basically accounted for in two ways (adapted from Morroni 1992: 158-162, 1994):

1. Amongst the technical factors, it may consider:
   1.1. The economies deriving from the capacity of some funds to operate on several different elementary processes at the same time. There is

[13] With regard to increasing scale (corporate growth) and the organizational and managerial changes, see Langlois (1989), Chandler (1990) and Chandler et al. (1997).

[14] For production functions with more than one factor it is assumed they are homothetic. This assumption could be placed in doubt as a real point of reference.

no doubt that the most singular case of such technical economies of scale is generated by the geometric perimeter/capacity ratio of vessels, tanks and containers: the three dimensional nature of space accounts for the fact that the recipient capacity grows at a greater rate than its cost. Indeed, cost increases bear an approximate relationship to its perimeter. Albeit that the walls of the recipient may need to be carefully lined, the cost of installation and operation will likely increase at a lower rate than its capacity. This may be measured in terms of the number of simultaneous elementary processes or units of volume (litres, cubic metres). More generally, this kind of economy of scale also occurs when, *due to technical improvements*, the number of the output-in-process units that a fund may simultaneously operate on, expands more rapidly than the increase in installation and operating costs. The relative reduction of costs deriving from funds capable of acting in several production lines at the same time (or, many more elementary processes) is also known as parallel-series scale economies.[15]

1.2. The indivisibility of some *phases* of the production process. On the one hand, economies deriving from the fact that the independently set costs of one or more phases could be divided by a lot of output units are here included. A typical case in point is the printing of books or magazines: the cost of preparing the first copy is fixed no matter how many units will eventually be printed. On the other hand, certain technical improvements may increase the working *rate* of fixed funds involved in a particular phase of a production process, without real damage to costs. It would perhaps increase the required number of lines in order to ensure that the extra capacity of improved funds is fully taken up. If the rate of activity is given by $\delta' = \delta/n$, δ being the irreducible rate, n will then be the total number of lines to be deployed. For optimum accommodation, n should be a positive integer number. If this is not the case, imbalances will incentivate the introduction of improvements to accelerate the remaining phases of the process.[16]

---

[15] These are the so-called parallel-series scale economies, as established by Leijonhufvud (1989: 213-4). In the nomenclature coined by Georgescu-Roegen they are referred to as *line parallel processes scale economies*.

[16] It is obvious that a fund may simultaneously gain in both its multi-capacity and the speed of its productive cycle.

2. An increase in the volume of operation is accompanied by a less than proportional expansion of inventories. The lower amount of input flows to be stored, the greater the output manufactured and sold.

Many authors add other economies of scale of monetary character to those given above. They derive from the prevailing dominant position of a firm in a sector. An advantage presumably associated with its big size. This leading position permits it to obtain better conditions for purchasing raw materials and intermediate goods, obtaining finance, etc. As those economies of scale do not derive directly from production, they are not studied herein.

In the case of technical economies of scale, it is important not to confuse the physical indivisibility of funds and cost economies caused by certain technical features that can improve their functionality. For example, a pipeline is a non-breakable fund. If it is shortened, it will no longer serve as a gas or liquid transport system. In this respect, it would be an indivisible fund. However, in these pages, indivisibility has been preferably defined in terms of operational capacity instead of physical integrity. There are therefore two different aspects to be considered: firstly, all pipelines of whatever diameters are physically indivisible as a whole and, secondly, each of them benefits from the volume/surface ratio and its impact on cost. Indeed, with expanding diameter the number of elementary processes to be carried out increases more rapidly than installation costs.

At any moment the production of a given good may be executed at different scales. Therefore there is a family of average cost curves reflecting the productive technique associated with each scale. An enveloping curve can be defined to encompass such average total costs curves. This envelope has been drawn at the bottom part of figure 69.

The enveloping scale curve (thick line) is not continuous because at any given moment in time, there are always a finite number of production techniques available. Each of these techniques is associated with a specific level of output per unit of time. Consequently, the full length of this enveloping curve is not necessarily always that of the minimum cost. In the example shown, average total costs of techniques 1, 2 and 3 are progressively lower due to economies of scale. But, out-with the normal production levels, narrow segments of ATC curves could be found whose cost level is higher than those of the nearest small scale. Moreover, as can be seen in the figure, the higher the level of production (relative position of the TC curves) the greater the impact of fixed costs.

A dynamic outlook would focus on the movement of the enveloping curves. The hypothesis to be considered would be that, on comparing the curves of two moments at sufficiently distant time, the effect (displacement) of starting-up new production techniques would be observed. This dynamic does not necessarily exclude any of the previously existing scales of the process.

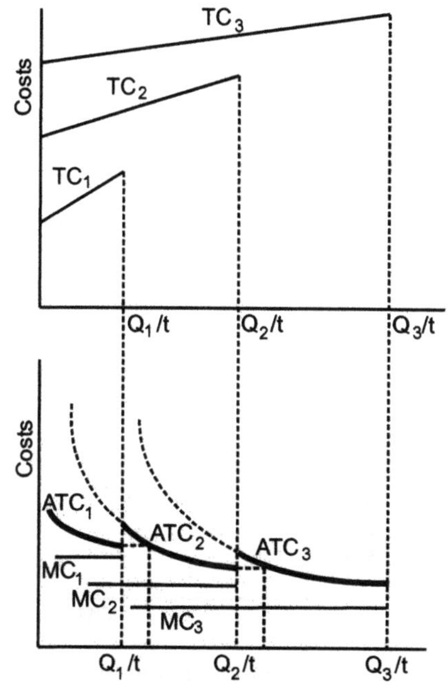

Fig. 69. Scale dimension curve

Prior to closing this section, the average total cost curve for a process working with multiple capacity funds should be defined. In such a case, a new cost element should be taken into account: the annual charge for depreciation corresponding to each kind of multiple capacity fund (A_j).[17] If α_j denotes the charge per hour, the expression of the average total cost is then given by,

$$ATC = \frac{\lambda}{n} \cdot \frac{1}{MCM}\left[c^{*}\left(\alpha + w + \alpha_j\right) + \sum_k f_k p_k\right]$$

where $n>1$, $MCM>1$ and $0<\lambda<1$

In the expression, factor Λ has been eliminated as it simultaneously influences both total cost and the level of production. It can also be observed

[17] For the sake of simplicity, it is assumed that the rate of depreciation is the same for all the j classes of multiple capacity funds.

that, the greater the capacity of the funds, the lower the average total cost due to the relative increasing weight of the factor 1/mcm. Ceteris paribus, the relationship between average total cost and the degree of the indivisibility of the funds (measured by MCM) is,

$$ATC = Z\frac{1}{MCM}$$

Z being the term that groups together the remaining parameterised components of the cost expression. Graphically, if the vertical axis represents average total costs and the horizontal axis the degree of indivisibility (or MCM), the hypothetical relationship between both variables would be as represented in figure 70.

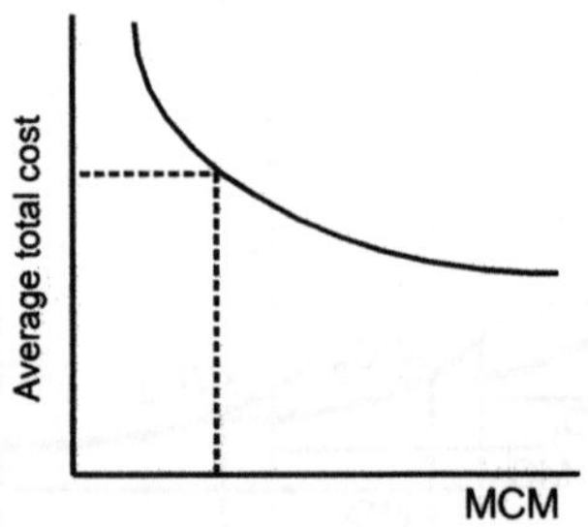

Fig. 70. Average total cost and multiple capacity funds

This figure attempts to show how the increasing multiple capacity of the funds will be translated into reductions in average total cost. Given that fixed costs may be assumed to changes equal, or more than directly proportional, to the changes of degree of indivisibility, it is plausible for the curve to fall slightly. To all events, it must be acknowledged that its specific pathway will depend on how technical innovation will affect the degree of funds indivisibility. It is obviously hard to make a specific prognosis with respect to this. All that is certain is that, sooner or later, the old plants will be given up.[18]

---

18 It must be remembered that the above analysis is limited to production activity. Therefore, general management overheads are not taken into account and, consequently, economies of scale associated with management improvements are ignored. This type of scale economies could perhaps be incorporated by an appropriate manipulation of the factor Λ.

6.6 Cost and flexibility

The line deployment of an elementary process enhances process efficiency. However, line processes suffer from a significant loss of flexibility: the ability to adjust to any change in the level and/or composition of demand is largely lost, and the costs of reprogramming the process are very high. In effect, the rigorous control of workers' tasks, the use of highly specialised funds and the improvement of inflow conveyor systems contribute to working time reduction per product unit exiting form the line. There is also interest in diminishing the rate of inflows per output unit, along with benefiting from economies of scale associated with the indivisibility of some of the funds. Given that line deployment represents a handicap with respect to flexibility, different strategies have been developed to attempt to offset this problem. These strategies are the subject of this section.

To start with, the greater the range of products a production unit is able to produce at a satisfactory average total cost, the greater its flexibility. This offers the establishment a competitive edge as its production may adapt to the variability of the demand while also opening up new market segments (Morroni 1992: 165). Therefore, the aim of implementing more versatile processes is to neutralise, albeit it partially, the uncertainty of frequently changing markets (Morroni 1991: 68). To achieve this goal, factors such as managerial skills and/or technical features of the process should concur (Jordan and Graves 1995). In any case, it is clear that an artisan production system, in which the human work fund predominates, is more flexible than a mass production system with highly specialised machinery. Therefore, there is a trade-off between economies of scale and flexibility (Dosi 1988: 1153ff.).

Unfortunately, the notion of flexibility has come into vogue in recent years. Although this has maybe blurred its meaning, the microeconomic production analysis has developed a clear scheme of the main forms of flexibility. This is shown in figure 71 (adapted from Morroni 1991: 69, 1992: 167). Time, understood as the logical short and long-term, has been the criterion used to classify the different forms of flexibility. The point is to draw attention to changes in the installed productive infrastructure or the use of what already exists.

Strategic flexibility refers to the firm's ability to innovate or change its processes and product lines (Gerwin 1993). The so-called flexible manufacturing systems, which are designed to take advantage of new IT and robotic technologies, gives processes the sought after flexibility to manage demand fluctuations. This flexibility is achieved better than with the Fordist methods, because it can produce small batches at competitive costs.

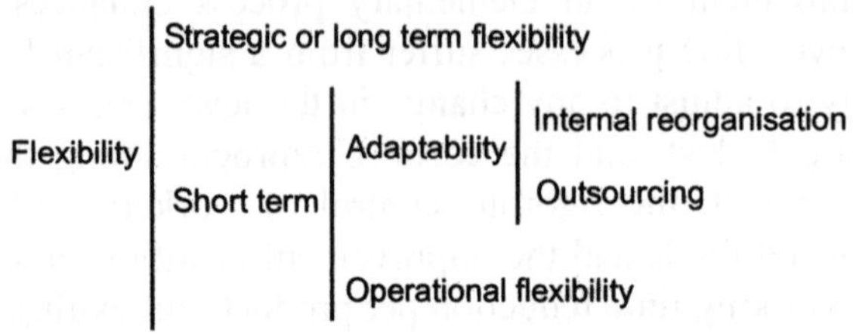

Fig. 71. Forms of flexibility

Short-term flexibility presents two very different cases. Firstly, if a plant is producing a single product, its level of adaptability is measured by the way average total costs react to changes in the amount of output per unit of time. Considering two plants operating at around the normal production level, the one endowed with greater adaptability will be the plant with lower increases in average total cost in expanding (or reducing) the output per unit of time.

Secondly, operational flexibility refers to establishments capable of generating a whole range of products. This type of flexibility deals with the (low) cost of changing batches or, what is the same, changing the proportions of the output-mix. This is a directly quantifiable problem of flexibility.

Short-term flexibility has two cases. On the one hand, given that this flexibility implies no change in the infrastructure already installed, the internal reorganisation of the production tasks may be the way to gain flexibility. This is the case, for example, of toyotism practices, such as using work-teams, internal outsourcing or worker recruitment on temporary time contracts (numerical flexibility).

The issue of flexible working times deserves closer attention. It comprises two main forms:

1. The extension of the standard working day with overtime or by increasing the number of shifts. The kind of adaptability involves longer working days, and weekend and night shifts. All of them have a negative effect on family and social life and even on the worker's welfare.
2. The availability of the services of the worker whenever required. In such a case, irregular working times appear. This distorts the organisation of the employees's private life as he/she should be permanently available

for any unforeseen needs the company may have. This situation is typical of female retailing workers.

All these forms of adaptability intensify the segmentation of the labour market.

On the other hand, outsourcing involves practices such as contracting companies for manufacturing components and spare parts or for providing cleaning, or safety services and so on (Morroni 1991: 73, 1992: 172).

Before finishing these paragraphs, the relationship between the size of a batch and the cost of storage is briefly taken into account. It may be expressed by the following:

1. Given a fixed cost of reprogramming (f) and a batch size (q_L), the larger the latter, the lower the ratio $F=f/q_L$,
2. The cost of storage (S) is assumed to be proportional to the batch size, i.e., $S=a \cdot q_L$.

This form of calculating the optimum batch size is the so-called *Standard economic order quantity* problem. In this simplified model, optimum batch size is considered a decreasing function of the changing cost of the batch. This means that the optimum batch size becomes smaller as the machinery used in the process becomes more flexible. This is represented in figure 72 (Piacentini 1997: 173-4).

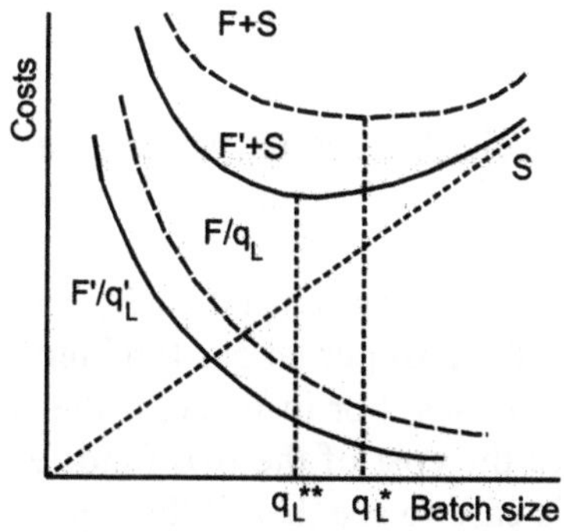

Fig. 72. Costs and the flexibility of the process

The minimum value of the sum $F+S=a \cdot q_L+f/q_L$ moves downwards and to the right as the reprogramming time is reduced. Indeed, this saving of time represents an increase in the degree of use of the (fixed) funds and, therefore, a lower charge for their depreciation. In addition, this saved time implies a lower need for storage. All of that gives rise to a reduction in the optimum batch size: from q^*_L to q^{**}_L. Smaller batches tend to be associated with the shortening of the production cycles and the reduction of the product in process and finished goods inventories.

6.7 Cost in services

In a previous chapter, services were classified into three different types. Leaving aside those that require a combination of complex skills and technical infrastructure, the other two types of services (support and consultancy, and entertainment) are characterised by the outstanding presence of the human work fund. As has already been indicated, it is very hard to standardise such services. For that reason, any assessment of the quality offered will have to tackle many subtle aspects. Alongside the leading role of the human work fund, the providing of such services requires the consumption of certain inflows, most particularly energy, and an appropriate physical space (a fund).

In order to simplify the cost expression, let us disregard the flows. Only costs for labour (wages) and the facilities occupied (rent paid or whatever financial burden) are incorporated.[19] In such a case, the average total costs for offering of the service may be defined as:

$$ATC = \frac{W+R}{H \cdot \frac{1}{T}} = T\frac{1}{H}(W+R) = T(w+\rho)$$

W and R being respectively the amount paid annually in wages and rent, T the duration of the service (elementary process) in hours, and H the amount of time of activity per annum (also measured in hours). Then, $w=W/H$ and $\rho=R/H$. This expression indicates, for each service rendered, the proportional part of the cost per labour fund unit employed and the cost of renting or leasing premises. For instance, if the service is an hour class, the expression provides the cost of the two funds employed per hour. It is assumed that $T=1$, so is ease to see that, if the remuneration paid (per hour) falls or rental charges drop, average total cost will likewise fall.

If the services considered belong to the area of commercial distribution, there is a new fund to be considered: the stock of goods purchased and sold. This fund is denoted Φ. It may be measured according to the average goods held in stores for the period of, for example, a year. The preceding cost expression may now be written as follows:

---

[19] As known, the fund-flow model is of partial equilibrium: value R is taken as exogenous to the model.

$$ATC = T\frac{1}{H}(W + R + \Phi) = T\,(w + \rho + \varphi)$$

If T=1 is defined as the length of the working day, that is, the hours the establishment is open, this expression allows an initial and indirect estimation of the technical stock rotation to be obtained. This is a key variable for improving the efficiency of commercial distribution. In effect, given that H is the number of hours of activity per annum, changes in φ would tend to suggest that average stock levels are falling. Ceteris paribus, this result would indicate a faster rotation of stock.

The importance of labour in the services considered herein explains the interest in controlling workers' tasks. It is important to ensure that employees do not go beyond what it is strictly necessary to provide a service. Work flexibility in order to ensure their availability when required is of crucial importance. Part-time contracts, temporary contracts to meet different needs, etc. are also very common.

The transport service should be considered on its own. This activity is a hard resource consumer. Moreover, it encompasses an enormous variety of services because of the different physical means used, the juridical nature of the supplier, or their defining technical features. Any cost model proposed will therefore be of a very limited scope. Let us take for instance the case of the road transport of merchandise. For the sake of simplicity, only the worker and truck funds will be considered, along with the flow of fuel consumed. The assumptions of the cost model are:

1. The output is measured in weight (q tons) per space (s kilometres): $q{\cdot}s$ or Tn.-km.
2. H indicates the annual hours of transportation activity.
3. The annual payments for the services of the driver are denoted W.
4. The element σ represents the annual cost for the use of the truck. Therefore, σ=A/H, A being the charge for depreciation and H the number of hours the truck is used.
5. To move each Tn.-km. energy inflow (fuel) is required. Its cost is $f_e \cdot p_e$, i.e., the technical coefficient of the litres of fuel consumed per Tn.-km. multiplied by its price.

Given the above elements, the expression of average total cost per Tn.-km. is given by:

$$ATC = \frac{W + \sigma + (q \cdot s)(f_e \cdot p_e)}{q \cdot s} = \frac{1}{q \cdot H \cdot \overline{v}}(W + \sigma) + f_e \cdot p_e$$

In this equation, the output $q \cdot s$ (Tn.-km.) has been separated as the outcome of three factors:

$$q \cdot H \cdot \bar{v}$$

$\bar{v}$ being the average velocity.

With regard to the prices of transport services, it must be pointed out that prices are usually set to achieve certain overall objectives with respect to income. Although in general terms, certain prices could be based on the cost of the service, as shown in the expression above, others may be established to generate revenue. In any case, the prices of transport services tend to be very specific to particular services, the price discrimination also being usual.

Finally, the costs incurred in the operation of telecommunications services are analysed. Following the explanation given in chapter 5, the cost expression for telecommunications services should underline the comparatively high importance of fixed funds ($j = 1, 2, \ldots, J$) for the network as a whole. There is no doubt that the investment in equipment and facilities required to transmit information represents by far the largest part of the financial expenditure of firms in this sector. Thereby the cost expression may be simplified, leaving out wages, maintenance and repair costs, and the consumption of inflows, such as power. To all events, it will be a first approximation. If required, all the cost factors may be quite easily integrated. Meanwhile, a simple expression of the cost per telephone call (the elementary process), would contain the terms (σ_j) and (ς_j). The first indicates the rental cost for each fixed fund[20] and the second, takes account of the number of elementary funds (or channels) into which each fund may be divided. Another cost factor is the H/T ratio: it indicates the number of elementary processes per hour. Given all these elements, the average total cost per call and channel, taking one hour as the unit of time, will be given by the following expression:

$$ATC = \frac{\sum_{j=1}^{j=J} \frac{\sigma_j}{\varsigma_j}}{H/T} = \frac{T}{H} \sum_{j=1}^{j=J} \frac{\sigma_j}{\varsigma_j}$$

In this very simplified equation, average total cost is equivalent to an optimised flow of call traffic. It is also assumed that all the fixed funds are

[20] The term σ_j indicates the charge per hour for the annual depreciation of each fund j. That is, the annual depreciation divided by 8,760 hours.

operative (to a greater or lesser extent) and such traffic is regular and sustained.

A typical problem in setting prices in the case of telecommunications services is that of allocating costs, especially those of the fixed funds shared by different services. These prices, in a dynamic sector such as telecoms, should also contemplate the possible evolution of the different segments of the market.

6.8 Costs in software production

In the preceding chapters, cost models have been established following the postulates of the fund-flow approach. This has been done both with respect to the production of traditional tangible goods (agricultural or industrial) or the provision of services. To some extent, this classification fails encompass the so-called information assets, of which IT applications are an outstanding exponent. Day by day, this group of goods is growing in importance in the developed economies. A common feature of these assets is that they are very costly to produce, which is in sharp contrast to the cost of reproduction. The former requires a huge amount of work hours to be expended in the product design, development and verification stages, whilst outlay for the latter is far lower, almost negligible. Historically speaking, this has always been the case of goods like books, newspapers or disks. In these cases, the cost of creation has always been very high, but their material replication very cheap. Obviously, this could not be applied prior to the development of the printing press, when books had to be copied by hand. The reproduction of manuscripts was a slow and costly procedure. Innovations made in printing and the advances in electromagnetic systems (to store and reproduce information) made prices plummet. As a result, such goods were massively distributed. Over recent decades a newcomer to this group of information goods has appeared: IT programmes and applications. As known, IT products are associated with the digitisation of all kinds of information contents. This technique has permitted the ease of reproduction by several orders of magnitude. Nonetheless, obtaining the first copy is still a labour intensive task. The expenditure in this first copy may be considered as sunk cost. In this and following sections, the most outstanding features of the production and reproduction of software are briefly described, in order to propose a preliminary model for the costs of IT products based on the fund-flow model.

As is known, IT is a complex technical system made up of material (hardware) and virtual (software) elements. These components of IT are incorporated into a large number of processes and products.[21] The earliest computers were used for mass data processing (finance sector, R+D, public administration). Their use then spread to the areas of corporate management, the leisure and entertainment industries, telecommunications networks, the control of systems of all kinds, and everywhere. Developments in IT systems have brought about a dramatic fall in the price of equipment, a rapid increase in storage and processing capacity and the ongoing improvement of the interface between machine and user.

Very briefly, the first computers were built and the transistor invented in the Bell laboratories in the 1940s. In the following decade, IBM became the leading player in the sector and the first prototype of an integrated circuit on a silicon sandwich base appeared. Over this whole period, the activity of the sector was sustained by funds from the American Federal Administration. With the arrival of the IBM 360 in the 1960s, the singular and strategic role of the software became apparent. Finally, the work of programming became a separate activity when, in the 1970s, the IT sector divided into three segments: the design and manufacture of semiconductors, the assembly of the hardware or computer equipment, and the production of software. The emergence of the latter sub-sector gave rise to the offer of standardised services in the areas of training and maintenance. This service already existed but was closely coupled with the selling of equipment and was understood as technical assistance. In the following decade, the massive spread of the use of PCs amongst the general public and the incorporation of IT components into all types of equipment led to a boom. At the same time, international outsourcing of computer assembly and software production operations emerged. Spectacular growth was maintained throughout the 1990s with the advent of the use of Internet and successive innovations in the telecommunications sector. This numerical revolution has led to a convergence of the telecommunications sector (telephone networks, fibre-optics, satellite communications, and so on), the IT sector (both hard and software, scanners, CD-ROMs, etc.) and the media sector (TV, cinema, radio, press). This confluence has given rise to the birth of a highly diverse and tentacular macro-sector: the ICT sector.

Examining the history of the IT sector in greater detail, the 1945-1965 period is now generally accepted as the first stage in the development of the industry. This was a time when programming was seen as part of the marketing and technical assistance services of selling the equipment. There was no intention of separately obtaining a return on the investment made

[21] The omnipresence of ITs justifies calling them *General Purpose Technologies*. This concept encompasses technologies with a wide-ranging impact, such as electrical power.

in software development. The programs were not protected by any kind of patent and the clients (large firms and the administration) were able to exchange them freely. However, given that such applications tended to be largely tailored to the needs of the individual client and to the specific architecture of the computer, such exchange was scarce. These customised programs were developed alongside the progressive introduction of mainframe computers.[22] It was at the end of the 1960s that independent companies were born to producing computer software (for mainframe computers). These software firms remained dependent on the large manufacturers, given that the software was sold jointly with the hardware. The market success of these emerging software ventures depended on their ability to correctly anticipate costs and properly manage the different tasks involved, while advertising was still of very little relevance. Nonetheless, in 1969 IBM initiated the practice of *unbundling,* i.e., selling the equipment, software, technical assistance and training all separately.[23] Progressively artisan methods gave way to specialisation in the design (what to do) and development (how to do it) procedures involved in software production. At the time there were, however, only a handful of people who could truly be considered programming experts. As expected, the knowledge accumulated and the algorithms created by the developers became the company's key factor to gain a competitive edge in software production.

Initially, programs were written in machine code, with an upper limit of 10 thousand instructions per packet. From the middle of the 1950s, the military and aerospace projects (SAGE and SABRE) required programmes containing a great many lines of code, making it essential to have some kind of programming language. FORTRAN appeared in 1957 as the result of two and a half year's work by a dozen programmers. It rapidly became

[22] This custom software was commissioned under contract with an open-ended quote –with the option of supplementary contracts– or at a set price, including a profit margin of between 8-15%.

[23] In 1967 the US Department of Justice filed an Antitrust suit against IBM. One of the reasons was the practice of packaging together the hardware, software and extra services for the same price. It was seen as insufficiently transparent, and as a tool with which to gain market share. As a result IBM, started to establish separate prices for systems engineering, training, maintenance, programming services and programs. It decided to split the software into operating systems (which include programming tools) and applications. Up to that time both had been supplied by the makers of the machine. The operating system or basic software market was not unbundled until the end of the 1970s. The software sector was atomised up to the end of the 1980s, when a few companies realised that it was the commercial, and not the technical aspects, that were far more decisive to gain market share.

the standard for scientific and technical applications. In June 1960, the programming language COBOL was presented. It was developed as the result of a contract with the US Defence Department. Both dominated two thirds of the programming of applications throughout the 1960s and 70s. Their use reduced the costs of writing the programs and simplified the learning process for new programmers (Campbell-Kelly 2003: chap. 2).

The years between 1965 and 1978 represent the second stage in the development of the software industry. It was in this period that the large hardware manufacturers separated the production of hardware and software. This led to an explosion in the number of programming firms. This was helped by the spread of the use of the *minicomputer* and of the standard IBM 360. At the time, the leading packages were applications dealing with inventories or payroll management. The biggest sellers were programs such as Autoflow and Mark IV from Applied Data Research (in 1965) and Informatics (in 1967), respectively. Such corporate software was marketed as if it were a capital asset. The greatest importance was therefore attached to live demonstrations for potential clients. It must be added that these programs required arduous training to be able to operate them. For this reason, the salesperson would offer training, documentation and after-sales services.

The increased functional capacity of programmes brought with it a more complex internal structure, a greater number of lines of code (by a growth factor of ten every five years) and an increase in the relative importance of the cost of software in overall IT investment (from 10% in 1960 to 40% in 1965). It was also shown that the cost of testing was greater than that of the application writing. However, no consensus was reached regarding the productivity and cost per line of programming. The estimates varied greatly: between one hundred and one thousand instructions per programmer per month and between one and ten dollars per line, respectively (Campbell-Kelly 2003: 102ff.). Finally, it should be pointed out that, from the end of the sixties onwards, people began aware that the potential profits were far greater from standard, rather than tailored, software packages. However, the technology on the hardware side had yet to advance sufficiently.

At this time, the debate regarding the legal protection of software started. The US declared in 1964 that software could be protected by copyright. However, the sector rejected the idea of software being considered a literary work. The main interest of the sector was in protecting the algorithms and know-how incorporated in the program. Meanwhile, the sector had opted to sell or rent out programs in machine language, granting a licence in perpetuity which could be revoked in the case of any unauthorised distribution of the program. Copyright only protected the supplementary written material. It was in 1967 that the patent was finally considered the best legal way to protect software. The first patents were granted in the United States for Autoflow, and for Mark IV in Canada and the United Kingdom.

The third period runs from 1978 to 1993. These years correspond to the boom in the use of the personal computer and the birth of the very first mass market for software. With a potential market of millions of machines whose owners all have similar needs, millions of sales of standardized shrink-wrapped software would be expected.[24] Therefore the programming sector began to develop the typical economic features of the publishing and audiovisual industries: the great comparative weight of the sunk costs of producing the first copy, followed by the negligible cost of reproduction, along with huge expenditure on promotion and advertising. It did, however, display a significant difference: the switching costs for package changing were very high, which opened up the possibility of taken advantage of planned obsolescence.

Between 1980 and 1995, the software industry was divided into operating systems, data base management systems, monitor control programs, programming tools and applications. In these years, the software sector grew at an average exponential rate of over 40% per annum in real terms. Most standard program developers were new start-ups whose commercial success was usually built around a main product. The growth of these firms was limited as their own particular segment of the market became saturated. Diversification came by way of acquiring out of house software, thus growing segment by segment – an approach which involved less risk than internal development.

In August 1981, the IBM PC architecture hit the market. Given that, at that time, microcomputers were running on different standards, it would take the PC several years to attain a dominant position. There is no doubt that the fact that the company decision to open up PC architecture facilitated its being copied, and soon the market was flooded with very competitively priced clones. Large and small firms alike started assembling computers and producing accessories.[25] Meanwhile Intel and Microsoft

---

[24] It should be pointed out that the personal computer had already been envisaged back in the 1960s, but neither the capacity nor the cost of the hardware and software of the time made it viable. It was considered a highly uncertain market and even as late as 1983, word processing was considered as the only important application for personal or home use.

[25] The manufacture of computers follows the pattern of a line process. The main phases are the production of the monocrystalline silicon ingots; the printing of the transistors, devices and interconnections of the circuit design by way of

specialised in producing chips and software, respectively (Freeman and Louçã 2002: 317-8).

The final period starts in 1993. In terms of hardware, these years are characterised by a rapid succession of generations of microprocessors, with extraordinary leaps forward in terms of performance. The microprocessor manufacturers enjoy a position of supremacy because the entry barriers are unattainably high. In the case of software, the confluence of significant economies of scale and network externalities has gives rise to a new form of competitive dynamic: the winner takes all. This will be discussed in detail below.

At the beginning of the considered period, different firms battled out to be the first mover in setting the standards for personal IT equipment. As standards define the segments of the IT market, once established, the company that sets them will enjoy an enormous strategic advantage. Once standards are set, the principal way to compete is to offer products or services at lower prices. In their struggle to establish their own standards, IT firms may adopt one of two general strategies: control or openness. The former is associated with a strong undisputed position. The latter pursues setting the standard by events, that is, because of the number of users has been increasing at a high rate. This strategy means avoiding the slow distribution of one's standards by way of individually negotiated licences. In order to expand the market rapidly, firms decide to reward the computer assemblers for pre-loading their own software while, at the same time, demanding they do not use any rival technology. Doing that, market penetration is quicker and a critical mass of users can be established as soon as possible. This is sufficient to persuade other users to adopt the same standard since it offers the best guarantee of all round compatibility. This strategy is accompanied by massive advertising campaigns and highly competitive promotion prices. Using these methods, the software sector has attained an unprecedented level of concentration.

Secondly, over recent years the general interconnection of computers (Internet) has been achieved. The spread of the world web wide has been

---

photolithographic exposure; the assembly of individual chips with connections of the input/output terminals and their encapsulation into a ceramic or plastic casing; the insertion and soldering of the chips onto a previously die-cut and printed circuit board; and the racking of these and the fitting of permanent connectors and other auxiliary systems. The final output is a computer. This manufacturing process is similar to any other industrial assembly line, despite the impressiveness of the white rooms in which the process takes place, the sophistication and cost of the fixed assets used, the skill levels required of the operators employed and the demanding quality controls to be passed.

coupled to the development of specific languages and protocols. At the same time, the number of services offered over it, have mushroomed. With regard this, in the second half of the 1990's there was a huge number of set-ups. This was greatly facilitated by the kind of output they offered: a virtually pure service. Nonetheless, shortly afterwards, many disappeared without trace. On the contrary, others succeeded and were mostly swallowed up by the large telecommunications operators.

Thirdly, it is important to note that software has been incorporated into all sorts of devices. Millions of mobile phones, cameras, cars and so on carry programs in their ROM memories. At this point, a special mention should be made of electronic games. This sub-sector emerged in the 1970s. First it was focussed on machines for amusement arcades. At the end of the 1980s it moved into the area of home consoles for video games and shortly after to the PC. It must be noted that the average lifecycle of games is around three months. Therefore, a vast array of titles is required to maintain a financial balance. These games are often produced by independent programmers most of whom will receive their corresponding royalties and disappear.

Lastly, recent years have seen the comprehensive digitisation of all different kinds of information assets. Everything from the software products through to the most diverse audiovisual products has been digitised. With regard to this, it should be remembered that digital products are characterised by the following:

1. As said, the cost of the production of the first copy is high or very high. Nevertheless, the cost of reproduction is virtually zero and the quality of the copy is identical to that of the original. Moreover, copies can be made instantaneously.
2. Such digitised contents can be instantly available at any place and at any time.
3. Inter-operatibility: digitalised contents should be operative in any system and network. This requires an enormous effort to standardise along with the use of shared protocols.
4. Information is a knowledge asset. Therefore, firms must distribute trial copies and, thereafter, retain customers.
5. Security and confidentiality is a requirement of any digital data transfer.
6. Digital data transfer suffers a rather special kind of congestion: the ease and capacity of transmission encourages the indiscriminate transmission of vast amounts of data over the Net. Very often this information is of doubtful interest.
7. All digitised products may be versioned to create more and more market segments. There are many features that may potentially be manipulated

for versioning: controlling the delay in getting access to information (the shorter the delay the more one pays), the place of access (cinema, TV pay channel, free use at home), the intrinsic quality (e.g. the resolution of pictures), the speed of access, and so on. The enormous versatility of digital products implies that offer and pricing may be adjusted to the user profiles.

As is widely known, the elements emphasised are the underlying factors which govern the technological and economic trajectories of ICT technologies.

6.8.1 Costs of producing and reproducing software

As has been said, information is expensive to produce but very cheap to reproduce. With the new digitisation techniques, the cost of reproduction has been reduced still further, at least with respect to electromagnetic copying systems. The cost of producing software and information products is basically concentrated on making the first copy. Because the master copy has to be paid for prior to starting large-scale reproduction, its cost is a sunk cost. In fact, reproduction often begins when the first copy has been successfully finished. If copies are not sold at the required level, it is very hard to recover the money invested. However, if the product is successfully sold, the firm may collect significant profits since the cost of producing additional units (more copies) is virtually zero. IT products have a nil marginal cost of reproduction. The same is also practically true of certain physical elements such as electronic chips, CDs and so on. Indeed, their cost of production is absurdly low compared to the cost of conception and implementation of the infrastructure required for their manufacture. In general, their production involves highly automated systems. In all such cases, the model of costing proposed will consider the average total cost as a basic function of the first copy expenditures.

The design and development of software (and other information products) is a highly labour intensive task. However, the cost of successive versions will benefit from some of the developments of the earlier versions and from the store of knowledge accumulated by its creators. Therefore, the sunk cost of up-dating projects can be supposed to be comparatively lower. A breakdown of costs involved in the production and reproduction of digital products is shown in table 7.

The table shows that the average total cost of *reproduction* only comprises the cost of depreciation of the facilities required and their maintenance and repair. Such costs are of certain importance when the reproduction involves making successive copies of video tapes with hundreds of video recording machines operating in parallel, but are far less significant when an IT equipment repeatedly makes copies of a digital product. To all events, the part of the overall outlay represented by the cost of the flows (mainly power) is so low that it may be considered zero.

Table 7. Cost of (re)production of information goods

		Cost element	
		Flows	Funds
Type of cost	First copy	Energy, raw material, consumables	Human work and facilities
	Reproduction	*Insignificant*	Facilities, maintenance

According to the cost models already proposed in this book, the cost of the production and reproduction of information products may be described on the basis of the following assumptions:

1. The process comprises one single phase and the output (one copy) is single.
2. Annual payments for human work services are considered zero in the reproduction stage.[26]
3. The process of reproduction takes place in a plant with an annual depreciation given by A.
4. H^T denotes the number of hours per annum (8,760 hours). H is also the duration of annual productive activity, given that it is assumed that the fund is active day and night without rest.
5. The magnitudes are measured in monetary terms and refer to the accounting unit of time (a year).
6. The cost of the funds and flows required for the production of the first copy is a certain amount denoted by Γ.

The total cost will result from the sum of Γ and the cost attributable to reproduction. The average total cost of the process of (re)production is therefore given by the expression,

$$ATC = \frac{\Gamma}{H^T \cdot 1/c} + \frac{A}{H^T \cdot 1/c} = \omega + c \cdot \frac{A}{H^T} = \omega + c \cdot \sigma^*$$

γ being the unit allocation of the cost of the production of the first copy ($\omega = c \cdot \Gamma / H^T$) and c the average copy time required measured in fractions

26 Therefore, in order to simplify the analysis, the hours of maintenance and repair operations have not been included.

of hour.[27] Given that the expression refers to a single reproduced unit, c is also the time gap between the instants of starting the copy process of consecutive copies. Therefore, the level of production is $1/c$ copies per hour. For example, if each copy takes 6 minutes then $c=1/10$ and production per hour ($1/c$) is 10 copies. On the other hand, the term σ^* represents the imputation of depreciation cost per hour, assuming the process is active 24 hours a day, all year round. The expression obtained also makes it clear that any innovation affecting the cost of the first copy, for example a programming technique that accelerates the writing of new software, will also lower the average total cost of the product.

Having reached this point, it should be remembered that information products take advantage of high economies of scale. This is a clear example of the indivisibility of one or more of the phases of a given production process: because of the cost of this phase is set independently from the volume of production, the greater this amount, the more significant the economies of scale obtained.

As is evident, the process of reproduction may be easily replicated in parallel. To such ends, it suffices to have N reproduction units to copy the original ceaselessly. The average total cost is then given by the following expression,

$$ATC = \frac{\omega}{N} + c \cdot \sigma^*$$

given that the level of production per hour of the reproduction system has risen to $N \cdot 1/c$, i.e., the number that each machine produces per hour multiplied by the number of machines. The conclusion is that parallel reproduction reduces the relative weight of the first copy.

The analysis now focuses on the issue of costs and time efficiency. It is clear that there are two types of problem regarding efficiency with respect to the reproduction of digital goods. On the one hand, possible intervals of inactivity of the equipment caused by breakdowns (either mechanical or software related) and, on the other, the loss of time derived from a copying process that takes longer than is normal in other plants.

---

27 In this expression there is an implicit assumption: the cost of the first copy Γ should be divided by the total number of copies made, whatever the amount of time invested in their successive repetition. For the sake of simplicity, in the model it is assumed that reproduction will take place during the course of one single year, given that the cost is divided by the number of copies produced per annum. If the considered span of time of reproduction activity is different, all that would be required is a parameter (greater or lesser than 1) to qualify H^T.

If the coefficient of use of the reproduction equipment is defined as $\varepsilon = H/H^T$, with $0<\varepsilon\leq 1$ and being H the number of hours per annum during which the system is operative, the expression for the depreciation cost per hour (σ) becomes,

$$\sigma = \frac{\sigma^*}{\varepsilon} = \frac{A}{\varepsilon \cdot H^T}$$

This is the cost of depreciation of the equipment when it is not operative 8,760 hours per annum.

A lower level of comparative efficiency may be included in the cost expression using a suitable concept of saturation. It is defined as,

$$\mu = c^*/c$$

c^* being the duration of the reference copying process. As expected, c^* is assumed to be lower than c, $0<\mu\leq 1$.

Let us take a hypothetical case with the following values:

1. A=100,000 monetary units.
2. H^T=8,760 hours.
3. H=5,000 hours.
4. c=10 min.=1/6 hours.
5. c^* = 6 min. = 1/10 hours.

In this case, $\varepsilon = H/H^T = 5{,}000/8{,}760 = 0.5707$ and $\mu = c^*/c = 0.6$.
The cost of producing one unit working H hours at the faster speed ($1/c^*$) is,

$$ATC^* = c^* \cdot \sigma^* = \frac{1}{10} \cdot \frac{100{,}000}{8{,}760} = 1.14 \text{ monetary units per copy}$$

By contrast reproducing the digital product with the incidence of the mentioned two types of time inefficiency would mean a cost of,

$$ATC = c \cdot \sigma = \frac{1}{6} \cdot \frac{100{,}000}{5{,}000} = 3.33 \text{ monetary units per copy.}$$

Therefore it is clear that the cost difference between the two processes is given by the expression:

$$ATC - ATC^* = c^* \sigma^* \left(\frac{1}{\varepsilon \cdot \mu} - 1 \right)$$

The average total cost expression incorporating the parameters referring to inefficiency is:

$$ATC = \frac{c^* \cdot \sigma^*}{\varepsilon \cdot \mu}, \text{ with } 0<\varepsilon\leq 1 \text{ and } 0<\mu\leq 1$$

As known, the description of the relationship between the average total cost and the output produced per unit of time is based on the conventional division of fixed and variable costs. Because such a classification does not entirely match that of the funds and flows model, bearing in mind the assumptions mentioned above, the cost of reproduction at any given moment of time is given by:

$$ATC=AFC=FC/q=\frac{\sigma^*}{q}$$

In this expression, expenditures on inflows and wages are not considered. No reference is made to the cost of the first copy. The reason is that its effect on cost reproducing can only be precisely measured if time is included. In the expression, q refers to the quantity of copies produced per unit of time, given a certain rate of activity (c) and assuming that the equipment operates without rest. The lower q, the greater is the burden of fixed costs. The shape of the (average) total cost curve is shown in figure 73.

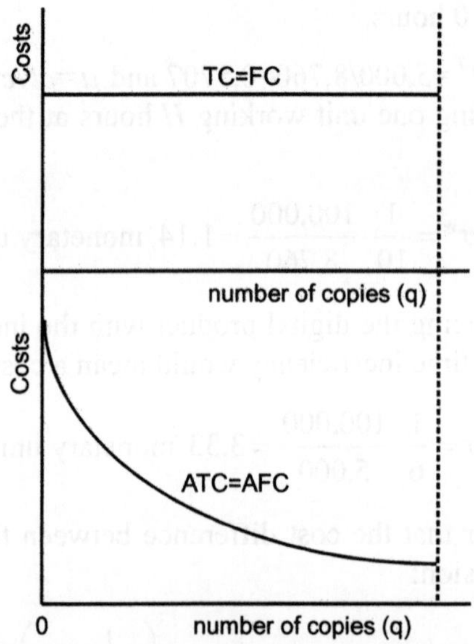

Fig. 73. Short-term plant cost curves in reproduction of information goods

The average curve falls and coincides with that of the average fixed costs. It is not feasible to operate beyond the point of maximum production, i.e., the maximum capacity of the reproduction system at any given moment in time.

With all these elements, a simple expression of the price of information goods could be given by,

$$p=(1+\chi)\cdot ATC=(1+\chi)\frac{\sigma^*}{q}$$

This price is the addition of reproduction costs and marketing and commercial distribution costs. O, in other words, ATC multiplied by the $(1+\chi)$ factor, χ being the promotional expenditures which are independent of the quantity sold.

6.8.2 Expanding the model of software costing

The static cost equations shown above constitute the basis of the production and reproduction of software. Some of the most salient points to be considered, if a more sophisticated version of the model proposed was undertaken, will be described below. As is evident, this advanced analysis should refer to the dynamic of costs and market prices of the software. In any event, these possible extensions of the model are left in the reader hands.

The first question is related to economies of scale. It was within the context of the Mark IV program (1960s) that the enormous economies of scale obtainable in the reproduction of software first became plain (Campbell-Kelly 2003: 128). People had heard of such economies in highly diverse sectors, but not in the case of IT activities. At the time experts still thought that the market price of software should be set at between ¼ and 1/5 of development costs. Therefore, the investment in written software could be recuperated with the first 4 or 5 sales. This outlook evidenced the existence of a very narrow market in which the cost of promotion and advertising would make very little sense. In this market, any additional services supplied to the purchasers should be charged separately. Only some years later did the size of the potential market reveal promotion and advertising as key factors for obtaining a return on investment. Indeed, from the beginning of the 1980s, it became common for developers to contract publishers to distribute draft versions of software applications. Publishers added attractive covering, related handbooks and promoted it. This relationship was similar to that of book publisher and authors (Campbell-Kelly 2003: 232ff.). In some cases, developers established direct agreements with the hardware manufacturers and computer assemblers in order to ensure their software came preloaded on new computers. In the mid-1980s, the promotional costs of a new software package for the professional users represented 35% of the total cost of the project. In PC applications, the percentage was even higher, given the persistent advertising campaigns thorough the mass media. Although benefiting from economies of scale required as large a market as possible, promotional costs grew as the market did. To obtain a return on such enormous investment, IT companies deployed diverse stratagems to further expand of the potential market for their software products. This marketing policy was decided within the context of a market with significant network externalities.

Another aspect to consider is the productivity of the process of designing and writing software packages. From its very beginning, there was a general awareness of the advances made in the production and capacity of the hardware, but opinions on software varied from cautious to pessimistic. Some estimation have quantified, the annual fall in the cost of hardware over the last decades at 35%, while for software, it is a mere 4% (Horn 2002: 47). Putting aside the quality raw data problem, this difference suggests a comparatively low increase in productivity in software production. Despite the fact that programs should tackle progressively more complex situations and the control of a larger number of peripherals, it seems clear there has been a productivity hindrance in writing software.

The conventional way for measuring productivity in software production is by the number of lines of source code produced per developer per unit of time.[28] Or, more specifically, the number of source code lines produced per hours-month. The rationale behind this indicator is that, although one program may be very different from another, all of them are written in lines of source code. Moreover, the number of hours per month is used because of it is very hard to determine the real length of a programmer's working day. Despite this rationale, the proposed indicator presents numerous conceptual and statistical problems. Therefore, its quantification usually gives rise to vastly differing results (Horn 2002: 74). There are many reasons for this: how to define a standard line of source code if looking at different programming languages, how to account for the direct ancillary activities required to compose the program, and so on.

Since the 1950s, there have been different generations of programming languages. These languages have been designed to facilitate programming:

---

[28] The source code is the content of a program that is understandable for humans. The object code is the same set of instructions but in machine language. A compiler translates the source code into object code. Purchasers of a packet have no access to the source code or programming lines.

they have led to a reduction in the number of lines of source code required to perform a particular function. Indeed, a given program that would have had one hundred thousand lines of source code in assembly language would only have 25,000 in PL/1 language, or 10 thousand if written in APL (Horn 2002: 76). A higher level language requires fewer man hours for an equivalent length of programming, since the developers require more or less the same time to write a line of programming whatever the language. Because there are fewer lines of source code per line expressed in machine code, measuring productivity by the number of lines is misleading. The fewer the lines of code required by a higher level language, the higher the programmer productivity. For the same function, high level language implies a fall in the cost of writing a program because it saves programming time.[29]

The indicator of the number of lines of programming produced per developer per unit of time would be appropriate provided that this task was a one-man job, always based on the same programming language, and producing similar results. Nonetheless, the difficulty of programming will depend more on the complexity of the program than its size (or number of lines of source code). Another problem arises from the fact that programs are developed by individuals who work constantly within the context of an interactive team. The greater is the complexity of the package to be developed, the more intensive is the communication between developers. With respect to the complexity of a program, it is tackled by conscientiously breaking it down into the smallest independent modules possible. Package architecture has been the most innovative area, along with the increasing levels of automation of the coding process. The importance of developing the concept has increased compared with the simple writing of lines of code. For that, the indicator of productivity mentioned above no longer makes sense. Finally, computer-aided software engineering has generated successive waves of innovation (structured programming, modular programming, object oriented programming) to facilitate the work of designing, writing and testing programs. These tools have facilitated the tasks of developing the concept, specification, internal documentation, integration, error correction, and so on of programs, as well as permitting the sharing of the same modules between different programs, all of which has contributed to an improvement in productivity throughout the entire process of software production, albeit one that is quantified precisely.

[29] According to different studies, the increase in productivity has been by a factor of between two and five if measured by the number of lines of object code contained (Horn 2002: 44).

However, writing a program has always been a somewhat risky venture. The reason is the bugs that may be encountered in them, whether in the development of the concept or in its execution. As known, bugs cause programs to behave in an unexpected way. Unfortunately, no data is available regarding the number of errors found in programs on sale. However, some studies estimate it to be between one and five serious errors per 1000 lines of source code. Some of these errors are hard to detect and debug. They may appear after years of use and be hard to isolate in the midst of a series of complex and interconnected systems. Moreover, attempting to correct one error may lead to others being introduced, given that corrections tend to make the structure of the program more fragile.

Apart from errors, users may complain that programs do not meet their expectations. Part of such dissatisfaction may be caused by the fact that the user is not sufficiently trained or skilled. Dissatisfaction could frequently be due to the programme appearing late on the market, as well as the use of new peripherals or capacities. However, the rapid obsolescence of software normally prevents the user from being aware of the quality and capacities of the program bought.

Any picture of the economic framework of software production also needs to be referred to conditions on the demand side. At this point, it is important to be aware that demand in the software market is driven by network externalities. Indeed, in certain areas of technological innovation, the simple fact that more and more people choose a given technical option implies that this will become the *best* alternative for any new user. If there is a consolidated user base for a given technical option, this will encourage new users to take it up. As is evident, guaranteeing compatibility with others is a key priority. This feeds a self-reinforcement mechanism (Arthur 2000). The resulting network will build up depending on the decisions made by the existing users and on expectations regarding the behaviour of newcomers. Such a positive feedback loop offers two kinds of advantages:

1. Direct benefits derived simply from obtaining access to the network: as the number of users increases, the amount of potential interconnections grows exponentially. So the value of joining a network will be progressively greater, the larger the number of individuals or agents already connected to it. To opt for a small or limited network means isolating oneself. In making such a decision, it is important to be aware of the risk of not being able to communicate with the vast majority.
2. The indirect benefits from belonging to the most extended network include, amongst others, a broader supply of technical assistance and more ancillary centre services.

The expansion of a user network generates a club effect.[30] Indeed, to start with, the club is too small to be appealing. At that moment, the rate of penetration amongst potential new users is very slow. Nevertheless, sustained growth in the number of members will enable it to pass beyond a certain threshold. Thereafter, the advantages of belonging to the club will become more and more explicit. As a result, the club will suddenly grow explosively. The positive feedback mechanism clicks in: given the number of users already connected it should believe that nothing will stop the mass adoption of the same technical option. The expectation of such massive affluence actually sustains it. Not long will pass before the chosen alternative will fill practically all the market. Despite the fact that the rest of options were on the market from its very outset, they will barely be able to scratch out a tiny little niche for themselves. They only collect a few survivors: those that have flatly refused to maintain an appropriate degree of complementarity. Sooner or later, this option is however doomed to fail. Nonetheless, congestion and trivia threaten the service quality of the winning network. If the capacity of the network is exceeded, users will receive poor service. Faced with the fact that it is not easy to find a less congested network, users will complain. With regard to trivia, all networks are predisposed to carry information which is banal or not pertinent. The greater is the number of users, the higher is the amount of spam to be found.

The mechanism triggered by network externalities is similar to a nuclear reaction or to a snowball rolling down hill. Once a certain level (or critical mass) has been surpassed, more and more people will join the group. This is a real phenomenon albeit that each person has his or her particular threshold to join the group. Ideological, ethical and other factors will only slow the process of mainstream incorporation. Critical mass mechanisms govern participation in social events and associations or groups of any kind (Schelling 1978: chap. 3). In such cases, on the one hand, there are individuals who will participate, regardless of the number of those who have already joined the group. These are people who will attempt to convince the others. On the other hand, there will normally be a group of individuals who will never participate, no matter what. Often these individuals prefer to be out of the club precisely because of such unanimity.

In processes of technological diffusion with network externalities, it is common that hazard occurrences have an exceeding influence on the course of events. The effect is so intense that practically nothing can be done to alter the course of events. This circumstance is a result of the in-

[30] Of course, not all clubs are shaped by network externalities. This is the case of items whose use does not hold strictly complementary relationships between them.

exorable self-reinforcement effect. What is decisive is not the fact that tips the balance, given that such events are often trivial, but the positive feedback thus triggered. In any case, a path-dependent contingency sequence is generated. Technology historians cite different examples of such processes: the case of the QWERTY keyboard or the MS-DOS operating system.

Towards the end of the 19th century the use of the typewriter became widespread (David 1985; Utterback 1994). The earliest machines had severe mechanical limitations. For that, special attention was paid to the design of models so such shortcomings were reduced. One area of attention was the layout of the keys on the keyboard. The first proposal, the QWERTY keyboard, was designed to limit the speed of typing since fast typing caused the levers and hammers to malfunction. Although the mechanical efficiency of typewriters soon improved, the QWERTY keyboard remained. Indeed, it is still used for computer keyboards today, even though nowadays there are no levers to move. Over the years, better layouts have been proposed, but none has been capable of displacing the QWERTY keyboard.

The explanation for the persistence of an imperfect design is to be sought in the process by which the typist learns. Although the QWERTY keyboard required more hours of training, once typing is mastered, the speed achieved using it is similar to that of any other keyboard layout. From 1880 onwards, the number of people trained on QWERTY grew steadily and therefore this layout gained force: nobody was interested in learning on others keyboards, nor were firms prepared to waste time and money on retraining staff. The cost of not having learnt QWERTY rose as the price of the typewriter fell. The use of this layout was extended further and became the de facto standard, whatever the technical considerations. The final outcome was not based on the intrinsic quality of the design, but grew from the strategic mimesis amongst the users.

The construction of a network with close relations of compatibility feeds the collateral phenomenon of lock-in. This is understood to be a situation in which users are dependent on a certain technical standard, because of the investment they have made in auxiliary systems and/or training.

Lock-in creates a captive market: after having opted for a particular technical standard, it is complex, risky and/or expensive to change. From that the user is held prisoner by the updates (or new conditions) that the manufacturers (suppliers) regularly make. For example, files generated by the latest update of a program will not be readable for earlier versions. It becomes compulsory to acquire the latest version. Table 8 shows the different types of lock-in and the costs associated with attempting to release oneself from its grip.

Table 8. Lock-in and switching costs

Lock-in types	Switching costs
Contractual deals	Claims and compensation
Equipment and software purchased	Investment in new equipment and software, cost of auxiliary devices
Learning	Learning costs, productivity lost after new training
Data bases and accumulated information	Cost of format conversion, risk of losing data
Specialized dealers	Find an adequate substitute
Fidelity	Loss of commercial premiums

The lock-in cycle starts with a given manufacturer or supplier hooking the potential customer with different marketing stratagems, all aimed at adding (and keeping her/him there) to its client base. At this point, it is important for firms for this base to gradually grow until achieving the threshold after which the self-reinforcement mechanism triggers. The period required to establish the customer base should be as short as possible. Different methods may be used. The most common is to distribute a very cheap or free copy sample (a demo). Another promotional stratagem is to sell computers with the software preinstalled. In this case, firms try to manipulate the expectations of the customers: that what is believed will become the standard will be the standard. If some packet is found on most of the machines that means this program is the standard. Among similar methods there is the so-called *vaporware*: an up-to-date version of a given packet is announced for the immediate future in order to dampen the sales of a competing product. This stratagem requires solemn declarations, such as "this will be the most popular program worldwide" and spectacular *mise-en-scene*. Doing that, an aura of inevitability is spread amongst users who distrust programs that are not very popular. Unfortunately, this race to be the first mover also bears the risk that the version sold may not yet be sufficiently mature and tested.

Software products are a system. Therefore an increase in value of one of the components will extend to the other remaining elements. This feature of package integration gives rise to what is called *bundling*: different programs are sold as if they were a single set which is offered at a price just below the sum of the separated prices of the individual components. Nobody will then purchase such applications separately, although some of them are scarcely used by users. Bundling is a form of versioning.

To summarise these particular demand conditions, it must be pointed out that the software market has grown out of the dynamic interaction of two main factors. Namely,

1. The huge economies of scale achieved in production and reproduction.
2. The impact of the network externalities or economies on demand. The value of becoming a member of the club of users increases faster than the expansion in the number of current users. Consequently, the network grows as a result of a particular self-fulfilling forecast: the expectation of success feeds this success.

However, there are other factors to be taken into account:

1. Setting the own technical standard (by deal or by de facto) is a key issue in controlling whatever emerging market. The first mover will gain an undoubted edge, despite the fact that continuous technical changes may lead to the disappearance of the current product standards and the associated productive infrastructure. Nevertheless, new innovations also open up new markets. The competition on standards is the competition *for* the market, not *into* the market. This explains the enormous importance of standards.
2. The exploitation of network economies is incentivated by the use of promotional stratagems aimed at rapidly surpassing the threshold number of users and built-in obsolescence. The advantage from the lock-in may be added, which is a form of constrained loyalty. And, finally, the blown up announcements of new and comparative better releases should not be forget.
3. Frequent business alliances, mergers and buyouts to anticipate the technology and market evolution.

A combination of economies of scale on the supply side and network economies on the demand side, accounts for the most spectacular of the economic phenomena related to the area of ICTs: the markets obeying the competitive dynamics principle of the winner takes all. Indeed, supply and demand conditions have been mutually sustaining and inclining the market as a whole towards the use of a principal single technology (supplied by a single firm). This dynamic has encouraged the emergence and consolidation of monopolistic positions at a speed and solidity never previously seen in the history of the market economy.

7 References

Abruzzi A (1965) The Production Process: Operating Characteristics. Management Science 11 (6): b98-b118

Ackello-Ogutu C, Paris Q, Williams WA (1985) Testing a von Liebig Crop Response Function against Polynomial Specifications. American Journal of Agricultural Economics 67 (4): 873-880

Amen M (2001) Heuristic methods for cost-oriented assembly line balancing: A comparison on solution quality and computing time. International Journal of Production Economics 69 (3): 255-264

Amendola M, Gaffard JL (1988) La dynamique économique de la innovation. Economica, Paris

Archibugi D (1988) Uso e abuso della specializzazione flessibile. Politica ed Economia 18 (9): 86-87

Arthur WB (2000) Increasing Returns and Path Dependence in the Economy. The University of Michigan Press, Ann Arbor

Atkinson AB, Stiglitz JE (1969) A New View of Technological Change. The Economic Journal LXXIX (315): 573-578

Babbage C (1969) Passages from the life of a philosopher. Augustus M. Kelley, New York. (1st edn. 1864)

._ (1971) On the Economy of Machinery and Manufactures. Augustus M. Kelley, New York. (4th edn. 1835)

Bailey EE, Friedlaender AF (1982) Market Structure and Multiproduct Industries. Journal of Economic Literature XX(9): 1024-1048

Barceló A (2003) Objetivo: cuantificar la reproducción. In: Mir P (ed) Producción, productividad y crecimiento (Economia i Empresa 12). Universitat de Lleida, Lleida, pp 9-36

Barceló A, Sánchez J (1988) Teoría económica de los bienes autorreproducibles (Libros de economía Oikos 29). Oikos-tau, Vilassar de Mar (Barcelona)

Baudin M (2002) Lean Assembly: The Nuts and Bolts of Making Assembly Operations Flow. Productivity Press, New York

Berg M (1980) The Machinery Question and the Making of Political Economy 1815-1848. Cambridge University Press, Cambridge

._ (1987) Charles Babbage (1791-1871). In: Eatwell J, Milgate M,Newman P (eds) The New Palgrave. A Dictionary of Economics. MacMillan, London, vol 1, pp 166-167

Best ME (1990) The New Competition. Institutions of Industrial Restructuring. Polity Press, Cambridge

Betancourt R, Clague CK (1981) Capital utilization. A theoretical and empirical analysis. Cambridge University Press, Cambridge

Birolo A (2001) Un'applicazione del modello «Fondo e Flussi» a uno studio di caso aziendale nel distretto calzaturiero della Riviera del Brenta. In: Tattara G (ed) Il piccolo che nasce dal grande. Le molteplici facce dei distretti industriali veneti. FrancoAngeli, Milano, pp 193-215

Boland MA, Foster KA, Preckel PV (1999) Nutrition and the Economics of Swine Management. Journal of Agricultural and Applied Economics, 31 (1): 83-96

Bonaiuti M (2001) La teoria bioeconomica. La "nuova economia" di Nicholas Georgescu-Roegen (Biblioteca di testi e studi 159). Carocci, Roma

Bresnahan TF, Ramey VA (1994) Output Fluctuations at the Plant Level. The Quarterly Journal of Economics CIX(3): 593-624

Bunge M (1998) Philosophy of Science. Transaction Pub., New Brunswich (NJ)

Campbell-Kelly M (2003) Form Airline Reservations to Sonic the Hedgehog. The MIT Press, Cambridge (Mass.)

Chamberlain EH (1948) Proportionality, divisibility and economies of scale. *The* Quartely Journal of Economics 1-2: 229-262

._ (1949) Reply to F. H. Hahn and A. N. McLeod. The Quartely Journal of Economics 1-2: 137-143

Chandler AD Jr (1990) Scale and Scope. The Dynamics of Industrial Capitalism. Harvard University Press, Cambridge (Mass.)

Chandler AD Jr, Amatori F, Hikino T (1997) (eds) Big Business and the Wealth of Nations. Cambridge University Press, Cambridge

Chenery HB (1949) Engineering Production Functions. The Quarterly Journal of Economics 63 (3-4): 507-531

Cobb CW, Douglas PH (1928) A Theory of Production. American Economic Review, March supplement: 139-165

Coriat B (1979) L'atelier et le chronomètre. Essai sur le taylorisme, le fordisme et la production de masse. Christian Bourgois Éditeur, Paris

._ (1990) L'atelier et le robot. Essai sur le fordisme et la production de masse à l'âge de l'électronique. Christian Bourgois Éditeur, Paris

Corsi M (1991) Division of Labour, Technical Change and Economic Growth. Avebury, Aldershot (UK)

._ (2005) The Theory of Economic Change: A Comparative Study of Marshall and the 'Classics'. Investigación económica LXIV (253): 15-42

Cusumano MA (1992) Shifting economies: From craft production to flexible systems and software factories. Research Policy 21 (5): 453-480

Daganzo CF (1997) Fundamentals of Transportation and Traffic Operations. Pergamon, New York

David PA (1985) Clio and the Economics of QWERTY. The American Economic Review 75 (2): 332-7

De Neufville R (1990) Applied Systems Analysis. Engineering Planing and Technology Management. McGraw-Hill, New York

Dosi G (1982) Technological Paradigms and Technological Trajectories: A Suggested Interpretation of the Determinants and Directions of Technological Change. Research Policy, 11 (3): 147-162

._ (1988) Sources, Procedures, and Microeconomic Effects of Innovation. Journal of Economic Literature, XXVI (3): 1120-71
Eichner AS (1986) Toward a New Economics. Essays in Post-keynesian and Institutionalist Theory. The MacMillan Press, London
Faudry D (1974) Difficultés d'estimation de la fonction de production micro-économique in agriculture. Économies et sociétés 5: 701-749
Ferguson CE (1969) The Neoclassical Theory of Production and Distribution. Cambridge University Press, Cambridge
Fisher FM (1971) Aggregate production functions and the explanation of wages. Review of Economics and Statistics 53 (4): 305-325
Freeman C, Soete L (1999) The Economics of Industrial Innovation, 3rd edn. The MIT Press, Cambridge (Mass.)
Freeman C, Louçã F (2002) As Time Goes By. From the Industrial Revolutions to the Information Revolution. Oxford University Press, Oxford
Frisch RAK (1965) Theory of Production. Rand McNally, Chicago
Fuss M, McFadden D (1978) Production economics: a dual approach to theory and applications. North-Holland, Amsterdam
Gadrey J (2000) The Characterization of Goods and Services: An Alternative Approach. Review of Income and Wealth 46 (3): 369-387
._ (2003) Socio-économie des services (Repères 369). La Découverte, Paris
Gaffard JL (1990) Économie industrielle et de l'innovation. Dalloz, Paris
._ (1994) Croissance et fluctuations économiques. Montchrestien, Paris
Georgescu-Roegen N (1966) Limitationality, Limitativeness and Economic Equilibrium. In: Ibídem Analytical Economics. Issues and Problems. Harvard University Press, Cambridge (Mass.), pp 338-355
._ (1969) Process in farming vs. process in manufacturing: A problem of balanced development. In: Papi U, Nunn C (eds) Economic problems of agriculture in Industrial Societies. IEA-MacMillan, London, pp 497-533
._ (1971) The Entropy Law and the Economic Process. Harvard University Press, Cambridge (Mass.)
._ (1972) Process Analysis and the Neoclasical Theory of Production. American Journal of Agricultural Economics 54 (2): 279-294
._ (1976) Energy and Economic Myths. Institutional and Analytical Economic Essays. Pergamon, New York
._ (1986) Man and Production. In: Baranzini M, Scazzieri R (eds) Foundations of Economics. Blackwell, New York, pp 247-279
._ (1988) The Interplay Between Institutional and Material Factors: The Problem and its Status. In: Kregel JA, Matzner E, Roncaglia A (eds) Barriers to Full Employment. MacMillan Press, London, pp 297-326
._ (1989) An Emigrant from a Developing Country: Auto-biographical Notes I. In: Kregel JA (ed) Recollections of Eminent Economists. MacMillan Press, London, vol 2, pp 99-127
._ (1990) Production Process and Dynamic Economics. In: Baranzini M, Scazzieri R (eds) The Economic Theory of Structure and Change. Cambridge University Press, Cambridge, pp 198-226

._ (1992) Nicholas Georgescu-Roegen about himself. In: Szenberg M (ed) Eminent Economists. Their Life Philosophies. Cambridge University Press, Cambridge, pp 128-160
Gerwin D (1993) Manufacturing Flexibility: A Strategic Perspective. Management Science 39(4): 395-410
Gibson JE, Mahmoud N (1990) The Moving Assembly Line. Real Labor Productivity Improvements Produced. Journal of Manufacturing and Operations Management 3(4): 293-307
Gold B (1981) Changing Perspectives on Size, Scale and Returns: An Interpretative Survey. Journal of Economic Literature XIX (5): 5-33
Grahl J (1994) Division of Labour. In: Areatis P, Sawyer M (eds) The Elgar Companion to Radical Political Economy. Edward Elgar, Aldershot (UK), pp 101-106
Hahn FH, McLeod AN (1949) Proportionality, Divisibility, and Economies of Scale: Two Comments. The Quartely Journal of Economics 1-2: 128-137
Heady EO, Dillon JL (1961) Agricultural Production Functions. Iowa State University Press, Ames
Heertje A (1977) Economics and Technical Change. Weidenfeld and Nicolson, London
Heizer J, Render B (1993) Production and Operations Management. Strategies and Tactics, 3rd edn. Allyn and Bacon, Needham Heights
Hennet JC (1999) Form the Agregate Plan to Lot-Sizing in Multi-level Production Planning. In: Brandimarte P, Villa A (eds) Modelling Manufacturing Systems. From Aggregate Planning to Real-Time Control. Springer, Heidelberg, pp 5-23
Heskett J (1980) Industrial Design. Thames and Hudson, London
Hicks J (1973) Capital and Time. A Neo-Austrian Theory. Oxford University Press, Oxford
Hill P (1977) On Goods and Services. The Review of Income and Wealth 23 (4): 315-338
Hodgson GM (1993) Economics and Evolution. Bringing Life Back Into Economics. Cambridge University Press, Cambridge
Horn F (2002) Les paradoxes de la productivité dans la production des logiciels. In: Djellal F, Gallouj F (eds) Nouvelle économie des services et innovation. L'Harmattan, Paris, pp 69-99
Ippolito RA (1977) The Division of Labor in the Firm. Economic Inquiry 15 (4): 469-492
Johansen L (1972) Production Functions. North-Holland, Amsterdam
Johnson GL (1970) Stress on Production Economics. In: Fox KA, Johnson GL (eds) Readings in the Economics of Agriculture. George Allen & Unwin, London
Jordan WC, Graves SC (1995) Principles on the Benefits of Manufacturing Process Flexibility: A Strategic Perspective. Management Science 41(4): 577-594
Kistner KP, Schumacher S, Steven M (1992) Hierarchical Production Planning in Group Technologies. In: Fandel G, Gulledge T, Jones A (eds) New Directions for Operations Research in Manufacturing. Springer-Verlag, Berlin, pp 60-74

Koopmans TC (1951) Analysis of Production as an Efficient Combination of Activities. In: Ibídem Activity Analysis of Production and Allocation. Proceedings of a Conference (Cowles Commission for Research in Economics 13). John Wiley & Sons, New York, pp 33-97

Kurz HD, Salvadori N (1995) Theory of production. A long-period analysis. Cambridge University Press, Cambridge

Kurz HD, Salvadori N (2003) Fund-flow versus flow-flow in production theory. Reflections on Georgescu-Roegen's Contribution. Journal of Economic Behavior and Organization 54 (4): 487-505

Lager C (1997) Treatment of Fixed Capital in the Sraffian Framework and in the Theory of Dynamic Input-Output Models. Economic Systems Research 9 (4): 357-373

._ (2000) Production, Prices and Time: a Comparison of Some Alternative Concepts. Economic Systems Research 12 (2): 231-253

Landesmann MA (1986) Conceptions of Technology and the Production Process. In: Baranzini M, Scazzieri R (eds) Foundations of Economics. Blackwell, London, pp 281-310

._ (1996a) Introduction. Production and economic dynamics. In: Ibídem (eds) Production and Economic Dynamics. Cambridge University Press, Cambridge, pp 1-30

._ (1996b) The production process: description and analysis. In: Ibídem (eds) Production and Economic Dynamics. Cambridge University Press, Cambridge, pp 191-228

Landesmann M, Scazzieri R (1996c) Forms of production organisation: the case of manufacturing processes. In: Ibídem (eds) Production & Economic Dynamics. Cambridge University Press, Cambridge, pp 252-303

Langlois RN (1989) Economic Change and the Boundaries of the Firm. In: Carlsson B (ed) Industrial Dynamics. Technological, Organizational and Structural Changes in Industries and Firms. Kluwer Academic, Boston, pp 85-107

Lavoie M (1992) Foundations of post-Keynesian Economic Analysis. Edward Elgar, Aldershot

._ (2004) L'économie postkeynésienne (Repères 384). La Découverte, Paris

Lee FS (1998) Post Keynesian Price Theory. Cambridge University Press, Cambridge

Leeuw F (1962) The concept of Capacity. Journal of the American Statistical Association 57: 826-840

Leijonhufvud A (1989) Capitalism and the factory system. In: Langlois RN (ed) Economics as a Process. Essays in the New Institutional Economics. Cambridge University Press, Cambridge, pp 203-223

Lerner J (1949) Constant Proportions, Fixed Plant and the Optimum Conditions of Production. The Quartely Journal of Economics 1-2: 361-370

Loasby BJ (1995) Book review: Mario Morroni, Production Process and Technical Change (Cambridge University Press, Cambridge, 1992). Structural Change and Economic Dynamics 6: 139-141

Maneschi A, Zamagni S (1997) Nicholas Georgescu-Roegen, 1906-1994. The Economic Journal CVII (5): 695-707

Marini G, Pannone A (1998) Network Production, Efficiency and Technological Options: Toward a New Dynamic Theory of Telecommunications. Economics of Innovation and New Technology 7 (3): 177-201

Marx K (1976) Capital. Penguin Books, vol. 1, Harmondsworth. (1st ed 1867)

Mendershausen H (1938) On the Significance of Professor Douglas' Production Function. Econometrica 6 (2): 143-153

McKay KN, Buzacott JA (1999) Adaptive Production Control in Modern Industries. In: Brandimarte P, Villa A (eds) Modelling Manufacturing Systems. From Aggregate Planning to Real-Time Control. Springer, Heidelberg, pp 193-215

Mir P, González J (2003) Fondos, flujos y tiempo. Un análisis microeconómico de los procesos productivos. Ariel, Barcelona

Mirowski P (1989) More Heat than Light. Economics as Social Physics, Physics as Nature's Economics. Cambridge University Press, Cambridge

Moriggia V, Morroni M (1993) Sistema automatico Kronos per l'analisi dei procesi produttivi. Caratteristiche generali e guida all'utilizzo nella ricerca applicata. Quaderni del Dipartimento di Matematica, Statistica, Informatica e Applicazioni 23. Universitá degli Studi de Bergamo, Bergamo

Morroni M (1991) Production Flexibility. In: Hodgson GM, Screpanti E (eds) Rethinking Economics. Markets, Technology and Economic Evolution. Edward Elgar, Aldershot, pp 68-80

._ (1992) Production Process and Technical Change. Cambridge University Press, Cambridge

._ (1994) Cost of Production. In: Arestis P, Sawyer M (eds) The Elgar Companion to Radical Political Economy. Edward Elgar, Aldershot, pp 52-57

._ (1996) Un'applicazione del modello fondi-flussi alla produzione di apparecchiature per telecomunicazioni. Università di Pisa, Pisa

._ (1999) Production and time: a flow-fund analysis. In: Mayumi K, Gowdy JM (eds) Bioeconomics and Sustainability. Essays in Honor of Nicholas Georgescu-Roegen. Edward Elgar, Aldershot, pp 194-228

Morroni M, Moriggia V (1995) Kronos Production Analyser©. Versione 2.1. Manuale d'uso. Edizioni ETS, Pisa

Mumford L (1963) Technics and Civilization. Harcourt, Brace and World, New York. (1st ed 1934)

Nell EJ (1993) Transformational Growth and Learning: Developing Craft Technology into Scientific Mass Production. In: Thomson R (ed) Learning and Technological Change. MacMillan Press, London, pp 217-242

Nolan P (1994) Fordism and post-Fordism". In: Arestis P, Sawyer M (eds) The Elgar Companion to Radical Political Economy. Edward Elgar, Aldershot, pp 162-167

Olhager J (1993) Manufacturing flexibility and profitability. International Journal of Production Economics 30-31: 67-78

Pannone A (2001) Accounting and Pricing for Telecommunications Industry: An Operational Approach. Industrial and Corporate Change 10 (2): 453-480

Paris Q, Knapp K (1989) Estimation of the von Liebig Response Functions. American Journal of Agricultural Economics 71 (1): 178-186

Paris Q (1992) The von Liebig Hypothesis. American Journal of Agricultural Economics 74 (4): 1019-1028

Pasinetti L (1981) Structural Change and Economic Growth: A Theoretical Essay on the Dynamics of the Wealth of Nations. Cambridge University Press, Cambridge

Petrocchi R, Zedde S (1990) Il costo di produzione nel modello "fondi e flussi". Clua Edizioni, Ancona

Piacentini P (1987) Costi ed efficienza in un modello di produzione a flusso lineare. Economia Politica 4 (3): 381-405

._ (1989) Coordinazione temporale ed efficienza produttiva. In: Zamagni S (ed) Le teorie economiche della produzione. Il Mulino, Bologna, pp 163-179

._ (1995) A time-explicit theory of production: Analytical and operational suggestions following a «fund-flow» approach. Structural Change and Economic Dynamics 6: 461-483

._ (1996) Processi produttivi ed analisi dei costi de produzione. In: Romagnoli A (ed) Teoria dei processi produttivi. Uno studio sull'unità tecnica di produzione (Collana di Economia-Serie Letture 11). G. Giappichelli, Torino, pp 91-106

._ (1997) Time-saving and innovative processes. In: Antonelli G., De Liso N (eds) Economics of Structural and Technological Change. Routledge, London, pp 170-183

Polidori P (1996) Aspetti tecnici nell'organizzazione dei processi produttivi agricoli. In: Romagnoli A (ed) Teoria dei processi produttivi. Uno studio sull'unità tecnica di produzione (Collana di Economia-Serie Letture 11). G. Giappichelli, Torino, pp 125-154

Pratten CF (1971) Economies of Scale in Manufacturing Industry. Cambridge University Press, Cambridge

Robinson J (1955) The Production Function. The Economic Journal CXV (1): 67-71

Romagnoli A (1996a) Agricultural forms of production organisation. In: Landesmann M, Scazzieri R (eds) Production & Economic Dynamics. Cambridge University Press, Cambridge, pp 229-251

._ (1996b) Caratteri generali dell'approcio analitico. In: Ibídem (ed) Teoria dei processi produttivi. Uno studio sull'unità tecnica di produzione. G. Giappichelli, Torino, pp 9-28

Rosenberg N (1994) Charles Babbage: pioneer economist. In: Ibídem Exploring the black box. Technology, economics, and history. Cambridge University Press, Cambridge, pp 24-46

Sabel C, Zeitlin J (eds) (1997) Worl of Possibilities. Flexibility and Mass Production in Western Industrialization. Cambridge University Press, Cambridge

Sandelin B (1976) On the Origin of the Cobb-Douglass Production Function. Economy and History XIX (2): 117-123

Scazzieri R (1981) Efficienza produttiva e livelli di attivitá. Il Mulino, Bologna

._ (1983) The Production Process: General Characteristics and Taxonomy. Rivista Internazionale di Scienze Economiche e Commerciali 30 (7): 597-611

._ (1993) A theory of production. Tasks, processes and technical practices. Clarendon Press, Oxford

Schelling TC (1978) Micromotives and Macrobehavior. W. W. Norton, New York

Scholl A (1999) Balancing and Sequencing of Assembly Lines, 2nd ed. Physica-Verlag, Heildelberg

Schumpeter JA (1954) History of the Economic Analysis. George Allen & Unwin, London

Shaikh A (1974) Laws of Production and Laws of Algebra: The Humbug Production Function, The Review of Economics and Statistics 56 (1): 115-120

._ (1980) Laws of Production and Laws of Algebra: Humbug II. In: Nell EJ (ed) Growth, Profits, and Property. Essays in the Revival of Political Economy. Cambridge University Press, Cambridge, pp 80-95

._ (1988) Humbug Production Function. In: Eatwell J, Milgate M, Newman P (eds) The New Palgrave. A Dictionary of Economics. MacMillan, London, vol 2, pp 992-995

Shephard RW (1970) Theory of Cost and Production Functions. Princeton University Press, Princeton

Shimokawa K (1998) From the Ford System to the Just-in-Time Production System. A historical Study of International Shifts in Automobile Production Systems, their Connection, and their Transformation. In: Tolliday S (ed) The Rise and Fall of Mass Production. Volume II (The International Library of Critical Writings in Business History 16). Edward Elgar, Aldershot, pp 93-115

Simon HA (1979) On parsimonious explanations of production relations. Scandinavian Journal of Economics 81 (4): 459-474

Smith CA (1971) Empirical Evidence of the Economies of Scale. In: Archibald CG (ed) The Theory of the Firm. Penguin Books, Harmondswoth

Spencer MS, Cox JF (1995) An analysis of the product-process matrix and repetitive manufacturing. International Journal of Production Research 33 (5): 1275-1294

Sraffa P (1926) The laws of return under competitive conditions. The Economic Journal, XXXVI: 535-50

._ (1960) Production of Commodities by Means of Commodities. Cambridge University Press, Cambridge

._ (1986) Sulle relazioni tra costo e quantità prodotta. In: Ibídem Saggi. Il Mulino, Bologna, pp 15-65. (1st ed 1925)

Steedman I (1984) L'importance empirique de la production jointe. In : Bidard C (ed) La production jointe. Nouveaux débats. Économica, Paris, pp 5-20

Stigler GJ (1991) Charles Babbage (1791+200=1991). Journal of Economic Literature XXIX: 1149-1152

Sylos Labini P (1980) Lezione di economia. Microeconomia. Ateneo, Roma

Tani P (1976) La rappresentazione analitica del processo di produzione: alcune premesse teoriche al problema del decentramento. Note Economiche IX (4-5): 124-141

._ (1986) Analisi microeconomica della produzione. Nuova Italia Scientifica, Roma

._ (1988) Flows, Funds, and Sectorial Interdependence in the Theory of Production. Political Economy. Studies in the Surplus Approach IV (1): 3-21

._ (1989) La rappresentazione della tecnologia produttiva nell'analisi microeconomica: problemi e recenti tendenze. In: Zamagni S (ed) Le teorie economiche della produzione. Il Mulino, Bologna, pp 19-49

._ (1993) Teoria della produzione. In: Zaghini E (ed) Economia matematica: equilibri e dinamica (Biblioteca dell'economista, Ottava serie II.1). UTET, Torino, pp 19-80

._ (1996) Variabilità degli elementi fondo e analisi intertemporale della produzione. In: Romagnoli A (ed) Teoria dei processi produttivi. Uno studio sull'unità tecnica di produzione (Collana di Economia-Serie Letture 11). G. Giappichelli, Torino, pp 107-122

Tinbergen J (1985) Technology and Production Functions. Rivista Internazionale de Scienze Economiche e Commerciali 32(2): 107-117

Utterback JM (1994) Mastering the Dynamics of Innovation: How Companies Can Seize Opportunities in the Face of Technological Change. Harvard Business School Press, Boston

Von Hippel E, Tyre MJ (1995) How learning by doing is done: problem identification in novel process equipment. Research Policy 24 (1): 1-12

Vonderembse MA, White GP (1991) Operation Management. Concepts, Methods, and Strategies, 2nd ed. West Publishing Company, St. Paul

Whitehead J.A (1990) Empirical Production Analysis and Optimal Technological Choice for Economists. A dynamic programming approach. Avebury, Aldershot

Wild R (1972) Mass-Production Management. The Design and Operation of Production Flow-Line Systems. John Wiley and Sons, London

._ (1989) Production and Operations Management. Principles and Techniques, 4th ed. Cassell, London

Wing C (1967) Evils of the Factory System demostrated by parliamentary evidence. Frank Cass and Company, London. (1st ed. 1835)

Winston GC (1974) The Theory of Capital Utilization and Idleness. Journal of Economic Literature 4 (12): 1301-1320

._ (1981) The Timing of Economic Activities. Cambridge University Press, Cambridge.

Wodopia FJ (1986) Time and Production: period vs. continuous analysis. In: Faber M (ed) Studies in Austrian Capital Theory, Investment and Time (Lecture Notes in Economics and Mathematical Systems 277). Springer-Verlag, Berlin, pp 186-194

Woodworth RC (1977) Agricultural Production Function Studies. In: Martin LR (ed) A Survey of Agricultural Economics Literature. University of Minnesota Press, Minneapolis, pp 128-154

Zamagni S (1982) Introduzione. In: Georegescu-Roegen N *Energia e miti economici*. Editore Boringhieri, Torino, pp 9-21

._ (1993) Economia Politica. Corso de microeconomia, 3rd ed. La Nuova Italia Scientifica, Roma

Ziliotti M (1979) Produzione e tempo: Il modello fondi-flussi di Georgescu-Roegen. Ricerche Economiche, 3-4: 622-649

Zuppiroli M (1990) Il «modello a fondi e flussi»: alcuni spunti interpretativi per la teoria dell'impresa agraria e l'economia agro-alimentare. Rivista di Economia Agraria, XLV (3): 527-547

Index

W

MIX
Papier aus verantwortungsvollen Quellen
Paper from responsible sources
FSC® C105338

FSC
www.fsc.org

If you have any concerns about our products,
you can contact us on
ProductSafety@springernature.com

In case Publisher is established outside the EU,
the EU authorized representative is:
Springer Nature Customer Service Center GmbH
Europaplatz 3, 69115 Heidelberg, Germany

Printed by Libri Plureos GmbH
in Hamburg, Germany